OUR ARCTIC YEAR

*During our arctic year our cabin was often snowbound, and we
needed snowshoes for our daily hikes in the surrounding wilderness.*

OUR ARCTIC YEAR

Vivian and Gil Staender

Photographs by Gil Staender

Volume 12, Number 4/1985
ALASKA GEOGRAPHIC®

The Alaska Geographic Society

To teach many more to better know and use our natural resources

Editor: Penny Rennick
Associate Editor: Kathy Doogan
Design and Sketches by Sharon Schumacher

ALASKA GEOGRAPHIC®, ISSN 0361-1353, is published quarterly by The Alaska Geographic Society, Anchorage, Alaska 99509-6057. Second-class postage paid in Edmonds, Washington 98020-3588. Printed in U.S.A. Copyright© 1985 by The Alaska Geographic Society. All rights reserved. Registered trademark; Alaska Geographic, ISSN 0361-1353; Key title Alaska Geographic.

THE ALASKA GEOGRAPHIC SOCIETY is a nonprofit organization exploring new frontiers of knowledge across the lands of the polar rim, learning how other men and other countries live in their Norths, putting the geography book back in the classroom, exploring new methods of teaching and learning — sharing in the excitement of discovery in man's wonderful new world north of 51°16'.

MEMBERS OF THE SOCIETY RECEIVE *Alaska Geographic®*, a quality magazine which devotes each quarterly issue to monographic in-depth coverage of a northern geographic region or resource-oriented subject.

MEMBERSHIP DUES in The Alaska Geographic Society are $30 per year; $34 to non-U.S. addresses. (Eighty percent of each year's dues is for a one-year subscription to *Alaska Geographic®*.) Order from The Alaska Geographic Society, Box 4EEE, Anchorage, Alaska 99509-6057; (907) 563-5100.

MATERIAL SOUGHT: The editors of *Alaska Geographic®* seek a wide variety of informative material on the lands north of 51°16' on geographic subjects — anything to do with resources and their uses (with heavy emphasis on quality color photography) — from Alaska, northern Canada, Siberia, Japan — all geographic areas that have a relationship to Alaska in a physical or economic sense. We do not want material done in excessive scientific terminology. A query to the editors is suggested. Payments are made for all material upon publication.

CHANGE OF ADDRESS: The post office does not automatically forward *Alaska Geographic®* when you move. To ensure continous service, notify us six weeks before moving. Send us your new address and zip code (and moving date), your old address and zip code, and if possible send a mailing label from a copy of *Alaska Geographic®*. Send this information to *Alaska Geographic®* Mailing Offices, 130 Second Avenue South, Edmonds, Washington 98020-3588.

MAILING LISTS: We have begun making our members' names and addresses available to carefully screened publications and companies whose products and activities might be of interest to you. If you would prefer not to receive such mailings, please so advise us, and include your mailing label (or your name and address if label is not available).

Acknowledgments

We would like to thank Jean Siddall of the United States Fish and Wildlife Service for her help in organizing the plant list for reference. In addition we appreciate the assistance of Mark Smith of the Oregon Museum of Science and Industry and Dennis Paulson of the Burke Museum, University of Washington, in organizing the bird lists according to the latest changes in nomenclature.

Vivian and Gil Staender

Another version of parts of *Our Arctic Year* appeared in the July, August, September and October 1981 issues of *ALASKA®* magazine.

Library of Congress cataloging in publication data:
Staender, Vivian and Gil.
  Our Arctic year.
  1. Natural history—Alaska—Brooks Range. 2. Wilderness survival—Alaska—Brooks Range. I. Staender, Gilbert. II. Title.
QH105.A4S726   1983   508.798'6   83-8815
ISBN 0-88240-176-9

*Dedicated
to
the memory
of
Adolph Murie,
Research Biologist
for the
United States Fish and Wildlife Service*

This quarterly issue of Alaska Geographic, "Our Arctic Year" by
Vivian and Gil Staender, deserves some extra explanation. It was
intended to be released simultaneously with a book store edition by
Alaska Northwest Publishing. Murphy's Law took over somewhere
along the line and "simultaneous" became a few months apart. The book
store edition was released several weeks ago, so it is possible some
members of Alaska Geographic Society may have purchased that issue
and for them this Alaska Geographic Society edition is a repeat.
If this has happened to any member, just let us know about it
and we will issue that member appropriate credit against any
other book purchase at the member's discretion.

Sincerely,

Robert A. Henning

President
Alaska Geographic Society

Contents

Ram Peak lay in the Brooks Range, near the site for our cabin in the wilderness.

Foreword

Ravens swishing their way across the western Brooks Range of Alaska have seen the tracks and smokes of Natives seemingly forever. In more recent times the snow and moss have shown the prints of others, the tracks at first advancing slowly as military exploration parties struggled on foot or pulled boats up swift rivers. A few decades later similar traces showed again as geological field parties crossed the region, their paths echoing the outline of the major rivers and the headwater passes. During the turn-of-the-century interval, raven eyes saw much activity with persistent clusters of sign as certain creeks grew cabins and sluices of short-term occupation. Though most of these fires went out and the walls crumbled with time, many did not disappear; old trappers, miners, or isolationists saw to that. Beginning several years before World War II, the ravens had to move over at times to let loud-droning planes share their skies. These craft turned winding valleys into straight-line routes, bringing mail and the annual supplies for the loners, hunters and scientists probing the country. The ravens began to note a few more chimney smokes and other man signs. Some faded with autumn solstice. Others, with their nearest neighbors at a distance of a hundred miles or more, persisted long enough for a life segment or a marriage to run its course. More and more summer visitors caught the dark birds' eyes as our nation's interest in environment soared in the late sixties. More signs persisted then as well. The sound trails of airplanes laced a network that was only partly seasonal. By the early seventies, bringing "d-2" and "pipeline," the rush was on. Trails and tracks, smokes and dust marked man's steady spread into these mountains.

Gil and Vivian Staender left quickly fading summer marks before the rush. Their trips are described in *Adventures with Arctic Wildlife*. Just before "d-2" became coloquial their chimney smoke appeared beside the river in the Brooks Range. They were not the only ones seeking solitude and time in these mountains, but "next door" was still well beyond most Outsiders'

*The Arrigetch Peaks
in the Brooks Range,
above the Arctic Circle.*

concept of neighbors. To many it would have been almost "next state over." They had scanned the land and found a near-blank space with earlier marks all but faded; more than 60 seasons had given time for slow subarctic recovery.

In this book the Staenders have shared their discoveries, their reactions to a year of isolation and the time to sense one's surroundings: the plants, animals, and weather. They tell of testing: testing the animals' acceptance of them; testing themselves both physically and mentally, learning how to meet conditions of life along the river. They absorbed much from the country, watching and recording. They found fear of bears, wonderment at close contact with animals at ease, deep satisfaction from true self-reliance, and a joy of space. They also provide illustrations of problems that can arise from planning errors and describe an unusually strong personal philosophy revolving around minimal disturbance. Their account of a year in the wilderness is a deep sharing of good and bad, of joy and strain. They have done their best to enrich their lives by drawing from this wild land in a way that touched the land lightly. Approaching the country in the way they did, I am sure that renewal won't require the more than 40 years of fading that followed turn-of-the-century miners just a few miles downstream. They withdrew a lot from the land's capital and yet hardly changed the ecological balance, leaving traces that were shallow on the land in comparison to their depth in the authors' lives.

Perhaps the decision of the authors will be followed by future users of this wild and humbling land who will reduce their impact even more by telling of wilderness in an all but nameless fashion. If place names are no more precise than regional, fewer of those who, unlike these authors, were not up to personal discovery would beat to death paths emblazoned on pages of winter reading while deluding themselves as to the wilderness nature of their experience.* Even when lightly sketched, track lines need some time to fade from ravens' view; northern tundra in particular will suffer from the insufficient healing time that results from over-specific publicity.

Read on. Accept the Staenders' open sharing; enjoy the country with them; and let their soft-stepping manner serve as an example to your own boots when you walk beneath the ravens' wings.

Frederick C. Dean

Frederick C. Dean
Professor of Wildlife Management
University of Alaska, Fairbanks

In deference to this point of view the editors have agreed with the authors to conceal the specifics of place names from the reader. "North Creek," "South Creek" and the big river are real rivers in the Brooks Range, but are concealed from overeager adventurers by the use of these pseudonyms. Other prominent geographic features, with the exception of the Arrigetch, are given temporary names also.

OUR ARCTIC YEAR

Introduction

You'll Never Come Back Alive!

In late August 1965 my husband Gil and I stood on the windy shore of an unnamed lake, on the Continental Divide, in the Brooks Range in arctic Alaska. After three months of isolation, working for the United States Fish and Wildlife Service, our food supply was nearly gone. Fall storms were already presaging winter, blanketing slopes all around with snow.

"What if our bush pilot forgot us here in the wilderness? Could we survive?" I asked my husband.

"We couldn't survive here." Gil looked around the bleak, windswept pass. "We'd have to move down to tree line and somehow build a shelter. Food would be the biggest problem. But I believe we could make it."

The idea grew. Could we survive on our own in the arctic wilderness for a year out of contact with civilization?

"Let's try it," I suggested to Gil a couple of years later. After four summers on Alaska's arctic tundra, we were entranced with that wild land, with its wide variety of plants and animals.

"Okay!" Gil's face shone. "Let's start planning." Our families and friends were apprehensive.

"You'll never come back alive."

"You must be out of your minds."

"How can a man spend a year alone with his wife?"

"If my wife and I did that, we'd kill each other."

Nevertheless Gil and I were determined to go ahead with our plan.

While I was finishing up the manuscript for our first book, *Adventures with Arctic Wildlife,* Gil took over the planning of food, clothing and equipment for our year in the arctic wilderness.

Beginning in the fall of 1968, Gil did a lot of research on arctic winter conditions. Books on every arctic exploration littered our living room. He received valuable assistance and suggestions from neighbors Calvin P. Burt and Frank Heyl, arctic survival experts who had worked as consultants for oil exploration parties on the North Slope of the Brooks Range. They had tried and

*A view of the big river valley
in which we settled
for our arctic year.*

3

tested clothing and equipment in extremely cold temperatures. He consulted them as to what was needed for adequate clothing in -50° weather, such as what kind of down clothing, heavy wool underwear and socks we should take.

Cal made mukluks for us, stressing the importance of avoiding waterproof material. He also designed and sewed some special curved mitts of elk hide, backed with beaver fur, to wear over our heavy wool mittens. Jean Siddall designed and sewed our big parkas, to be worn over thick down jackets.

Cal re-webbed some old Yukon model snowshoes for us to use. He also advised us on what stoves, lamps or lanterns would be suitable and lent us an antique Aladdin kerosene lamp. We avoided gasoline because it was more hazardous.

Gil and I had never built a log cabin, so Gil read books on the subject and there were long consultations with Frank and Cal on how to go about it and what tools we would need.

Our four previous summers living near the crest of the Brooks Range provided Gil with a base of information to begin food planning. He thought four times the amount of food we had taken in for an arctic summer would be adequate. If supplemented with moose meat and fish we should have more than ample provisions to see us through.

Most of the food was bought in the Portland area because it needed to be repackaged to save weight and space. Most packaging was plastic: bags, jugs, square containers, which were lightweight, rainproof and collapsible. We also took some large,

Frank Tobuk gave us a stovepipe oven to use when we built our cabin.

lightweight metal containers for rodent-proof storage.

Our family physician, Dr. Ray Moore, gave us some medical supplies and wise advice on how to handle some emergencies. Bernice Sitton, a medical technician and bacteriologist, prepared our first-aid kit.

Gil and I had become dissatisfied with urban living. In 1955 our new home was the only one in a tall fir forest, with a winding two-mile road into Lake Oswego. We were now surrounded with new homes and schools and the narrow winding road had been replaced with a four-lane street with speeding traffic, so it was difficult to enter it with our bicycles. The tall trees had been cut down to make room for new buildings.

The final push came when we learned that Mount Sylvania, a 600-acre estate a quarter of a mile away, was sold to a developer. The parklike estate, with many species of wildlife and native plants, was our favorite place for nature study. We learned that there would be 11,000 people living up there, probably with that many dogs and cats.

"There goes more wildlife habitat." Gil looked sad. We decided to do something about it.

"If we sell our home we can buy some land in central Oregon for a nature preserve to replace some of that being lost," he suggested. "I've always wanted to do that."

"It's getting too crowded here anyway," I added. "We arc living where there is the highest concentration of people in the state. It's time we moved."

Gil and I began looking for property in central Oregon. Jim Ramsey, a mutual friend of the owner of a 160-acre piece of unspoiled land enclosed by the Deschutes National Forest, suggested we look at it. After a tour of its centuries-old junipers and tall ponderosa pines, with bitterbrush and bunch grasses, we decided it should be saved and would be a suitable wildlife habitat for a nature preserve. So we sold our home in Lake Oswego and bought the 160-acre piece of forest land from Jack Shumway, who said he won it in a poker game. Gil and I immediately willed it to The Nature Conservancy, just in case we didn't "come back alive" from the arctic wilderness.

"We'll store your furniture for you," Nick and Kay Dodge offered. "Don't pay for storage." So we put our furniture and other possessions in their basement, sold our car and arranged to deliver our new pickup with canopy to a dealer in Fairbanks, who was buying it from a dealer in Portland. This would provide for the transportation of our supplies.

Gil bought a year's supply of film for three cameras and had them winterized before we left for the north.

1·On Our Own

July 20, 1969. After six months of intensive preparation and planning we were actually on our way. Our bush pilot was waiting for us and our 1,600 pounds of food and other supplies at Bettles, Alaska, on the Koyukuk River, 100 miles northwest of Fairbanks.

Gil got in beside the pilot, Daryl Morris, in the little Cessna 180. The small, four-passenger aircraft was jam-packed in back with bundles and boxes. At first I couldn't get into the space they had saved for me. After some rearranging, I managed to squeeze in between packages behind the seat, wedged in with my knees against my chin. The plane motor droned as we flew north over the subarctic wilderness. Winding rivers, bends accentuated by parallel curving lakes, lay between dark stubble of spruce. We zoomed over low barren hills, then rugged snowcapped mountains beckoned to the north as we passed the jagged Arrigetch peaks.

My brain was reeling with countless questions.

Were we taking too big a risk? What would the approaching winter do to us? How difficult would it be to build a cabin with simple hand tools? (We had never built a cabin!) Could we build the cabin before winter came? What is it like at 40° or 50° below zero? What would the psychological effect of the long winter night be on the two of us, alone together for so long? Would we get "cabin fever"? So much was unknown to us.

I was jolted back to the present as the little plane banked and turned. Both men were scrutinizing the meandering river below, searching for a good spot in which to land.

Gil pointed to a narrow gravel bar beside a low flat island covered with shoulder-high feltleaf willows. "That bar looks *almost* big enough to land on." A large bull moose lay on the island. He looked so peaceful there.

The pilot shook his head. "I don't like it. See all that driftwood scattered on the sand."

When the Cessna circled the moose lurched to his feet, splashed across a channel to another island and disappeared in the trees.

We looked north over river bars brightened by fall color.

"Let's look for a bigger landing spot." Daryl revved the motor and we headed upstream for a few miles. The men kept their eyes peeled for a likely landing spot. Nothing seemed to suit them. We headed back downstream and passed over the spot which we had circled before.

A large pale gray wolf crouched low and held still for a moment, looked up at the plane, then fled into the brush. One hundred feet away a porcupine scurried for cover.

"Maybe I could land there if we approached from another direction." Daryl swung the little plane around and we circled over the same bar again.

I couldn't straighten my legs, which were cramped from my doubled up position. A sickish feeling crept into my stomach.

"We'll try it," our pilot announced. He swung to the south, then we came around a bend in the river, heading north again, lost elevation, and prepared to land. Wedged in so tightly, I didn't need to brace myself. Tense, I waited for the wheels to hit and bounce on the bar.

Just then Daryl gunned the motor to gain elevation again. My mouth was dry as cotton.

"That bar isn't long enough," he remarked tersely. "Let's look for another."

"Whew!" Stoic Gil swallowed hard.

Again we flew around and around, then back and forth.

"Maybe you can try it from another angle," Gil put in, "or hit the bar farther out, actually in the water."

Moments later the wheels thumped the gravel, bounced up, then down again. We careened down the bar, scattering gravel, and ground to a halt at the far edge.

"Wow!" I gave Daryl a hug.

We all pitched in to unload our baggage. Then Daryl took off to bring us two more loads of supplies and helped carry the last bundles off our "airstrip."

"You don't need to come back for us," we told our pilot. "We'll come out on our own . . . a year from now." Gil and I planned to float down that arctic river in a rubber raft for 200 miles, to the first Indian-Eskimo village.

Our supplies heaped beside us, it was time to bid the pilot good-by.

"Be careful!" Daryl waved as he accelerated the motor.

"You be careful!" We yelled after him as he bounced down the beach until he was airborne. The hum of the plane motor faded into the distance.

"Sure is quiet here." Gil's voice was hushed.

"It seems unreal that we'll get no news of the outside world for a year." I looked at my husband. "Are you scared?"

He looked around. "Better check for grizzlies. Lots of bear tracks around here."

With binoculars we carefully scrutinized the slopes all around. Feltleaf and other shorter unidentified willows and dwarf birch crowded the riverbanks and lower slopes of the mountain on the opposite side of the river.

We felt small and insignificant as we stood on that gravel bar in the wilderness. Shoulder-high willows covered the center of the 100-yard-wide island. Many wolf tracks marked the sand, along with moose and bear tracks. About six mew gulls cruised over. Arctic terns hovered over shallow water at the head of the island. Redpolls called, "Chi-chi-chi," their cheerful flight song, as they flew over in a typical undulating finch pattern. We heard gray jays call to each other as they winged back and forth through the spruces across the river.

Gil erected our small sleeping tent, designed for car camping. This tent had seen yeoman's service on previous trips to the Brooks Range.

We felt a faint sprinkle of rain, so we covered supplies with a blue nylon tarp. The bright blue color seemed wrong here. In this wilderness man should blend into the environment.

We spent much time carefully surveying the slopes with binoculars.

Gil pointed to a prominent peak to the north, across the river. "That mountain resembles Mount Eisenhower in Banff Park."

"It certainly does. Let's call it Mount Ike."

We found it helpful to give temporary names to geographic features around our camps. Because we keep detailed data on the flora and fauna we need reference points.

We stacked all our bundles close around the tent, as a sort of barrier against marauding bears. Evening came. Suddenly we felt very tired now that the excitement and hurry were over. All we had to do now was to take care of ourselves for a year. There would be no way to call for help.

"There's a grizzly." Gil's voice rose with excitement. He pointed across the river. The huge, light brown beast ambled along the riverbank directly opposite. I'd forgotten how large Barren Ground grizzlies were. He climbed down and eased into the stream, then started swimming and floating downriver.

"He doesn't know we're here," Gil commented. We watched him through binoculars.

The bear's coat was shaggy. He looked thin and hungry. About 200 feet downstream the grizzly clambered up on the opposite bank again and traversed along the top, keeping close to the stream. He stopped to graze on some berry plants, nearly hidden by the birch and willows. Soon he was back in the river, swimming again, low in the water, with only the top of his head and back visible. Bruin seemed to enjoy the water. Where the stream was shallow he galloped, splashing in the water like a human youngster, still heading downriver. Later he climbed up on the bank and resumed his journey.

"He's free as the breeze, doing just what he wants."

"Listen." Gil put his finger to his lips. "Something big is in the brush behind us." Our hearts lodged in our throats. We stood rooted in position. What was it?

One hundred yards away, the lanky feltleaf willows parted and a big bull moose came into view. He wore a magnificent set of antlers, not fully developed, which were clothed in rich brown velvet. The moose strode our way until he saw us. He stared as if he couldn't believe his eyes, then wheeled and splashed toward the bend in the river. He stopped, turned and stood facing us spraddle-legged. Finally he swung around and trotted out of sight.

After a supper of canned sausage, bread and soup, we began building a ring of fires around our camp.

"We need lots of smoke. It might scare away any other bears that come by." Gil added gobs of wet moss, grass and leaves.

"That's a good idea." I collected soaked debris from around driftwood on the gravel bar. The more I thought of grizzlies roaming around in the night, the more soggy debris I added to our bonfires.

We felt lonely and thoroughly frightened as we got ready for bed. We began to have second thoughts about the wisdom of being completely alone for a year in the wilderness with no radio for communication with the outside world. If we had any serious injury it would be a long time before anyone knew of it. Several times during the night of half light we heard strange sounds outside the tent. We got up to check the area with binoculars, trying to imagine what was hidden in the brush, before we went back to restless sleep.

•

July 21. We awoke to the pungent smell of wood smoke. A south wind had carried a dull gray haze from a forest fire somewhere to the south.

We began our day with a thorough check for grizzlies. We had been terrified by our experiences with the Barren Ground grizzly the previous summer, when we had camped at the head of North Creek and were charged twice. We aimed to be extremely cautious.

"None in sight," Gil announced.

A family of gray jays called from spruce woods nearby south of our island. Several ravens squawked on the mountainside above.

The gravel bars were aglow with the rosy color of arctic fireweed. We anticipated a green salad from these plants, which were loaded with vitamin C. Creamy pallid paintbrush, purple asters and hawk's beard rosettes, like little green and gold doilies, added to the floral carpeting.

The first day we planned to explore and decide where we should build our survival cabin. We would take short exploratory trips and return often to camp to see that our supplies were safe. Should we lose our food and other supplies we would have the bitter disappointment of having to leave here and float out to civilization immediately.

To the north we saw tall pointed spires of spruce. Wading a two-foot-deep, narrow channel of the river, Gil, in hip boots, carried me across piggyback. I wore shoepacks. Where North Creek emptied into the big river we found the largest spruce in the vicinity.

With his rifle at the ready Gil entered the thick grove of trees, watching for bear. I followed close behind. The spruce trees were larger than we expected. This was a narrow peninsula, where

Hawk's beard rosettes, like little green and gold doilies, added to the floral carpeting.

alder, cottonwood and a few paper birch formed a thick underbrush. But we were concerned about the limited visibility. On previous trips to the arctic we were accustomed to camping on the open tundra, with a clear view all around. Gil and I returned to check our camp and see if our supplies were still intact.

After exploring in all directions, we decided that the spruce-covered peninsula at the mouth of North Creek was the best location for our cabin. We had time to move our six five-gallon cans of kerosene before supper. Gil carried two at a time, while I carried one, three-quarters of a mile to our selected site.

Early next morning we began the big job of moving all our possessions to the new location. We relayed the bundles about 150 yards at a time so we could stay close together, keeping supplies consolidated, with firearms at the ready at each relay station. Packages were carried in our arms to the first station at the end of our island. With my first load, I saw a wolf disappear into the willows across the stream. I whistled to Gil and called, "Wolf!" He came running. We saw it again about 200 yards away. The wolf was a typical gray color, with a dark streak along its back.

Gil carried me piggyback across the stream. We inspected the wolf tracks and found that there were two wolves traveling together. Their footprints led back across the stream toward our camp. While we slept last night the two canines had approached our tent. I wonder what they thought of human beings entering their territory.

While we worked at relaying bundles a couple of curious ravens flew low overhead, their wings making a swishing sound.

"Cronk," I called out to them. One answered, "Cronk." Each time I called it answered. If I soft-talked to it there was no reaction. I "cronked" again and it responded, "Cronk." Its companion was quiet. They flew away together.

Occasionally a mew gull or two flew over. Not far away a pair of arctic terns hovered over a shallow stream. How lovely these creatures were, even more beautiful than I remembered.

We noticed clouds and rain showers on the high peaks, toward the Arrigetch Mountains. Sprinkles of rain kept us hurrying. At 2:00 P.M. we were still working on the gravel bars of the big river and getting tired. But we did not rest. White butterflies fluttered lazily ahead of us while we trudged through bright patches of fireweed. We kept swiveling our heads around, like owls, on the lookout for bears.

Relaying through a thick willow patch we kept close together. Our arms ached as we continued without rest. By day's end our bundles were all together on the gravel bar below our chosen spot.

"My arms feel like they're two inches longer." My shoulder sockets ached as I scrambled up a five-foot embankment to where our kerosene cans were stacked. Gil hoisted the last things up to me as clouds of gnats and mosquitoes swarmed around our faces.

"Hurry! Break out the mosquito repellent!" I was frantic. We were tired and hungry . . . and scared.

We trudged through bright patches of fireweed.

*A grizzly trail —
right through the sand
near our camp!*

The limited visiblity in the trees worried us. Nervously we prepared our evening meal on the camp stove.

"Wow! Look at this!" Gil pointed out a staggered line of round depressions in the earth — a telltale sign.

"A grizzly trail — right through the middle of camp!"

There was an oft-quoted safety rule of the northern wilds, Never camp by a bear trail. Two large spruce trees, three feet apart, formed a portal for the path leading into the woods.

"Heck! This is a bear tree. See where they chewed the bark. Here's another bear tree. All around here, bear trees! Grizzlyville!" He plucked at blond grizzly hairs caught on the brown bark.

We inspected the bear scratching posts — in our camp. Grizzlies had stretched up to bite and claw the trees seven feet from the ground. We walked on the grizzly trail, waddling from side to side, stepping in the saucer-shaped depressions they had left.

We regarded the situation as serious as we set up the tent and crawled inside. Canvas walls do not deter a bear much, but, for some strange reason, we did feel safer inside.

In the wee hours we awoke to hear twigs snapping in the brush behind the tent. We got up to investigate in the dim light.

"There's a huge mammal moving near the bear trail." Gil kept his voice low. He picked up his rifle and watched through the "Portal Trees." Nothing in sight. But still we heard movement in the brush.

Hardly daring to breathe, we heard the animal moving softly away, down the bear trail.

"Must be a bear," Gil surmised. When we heard nothing more, we went back to bed, but lay listening for a long time, afraid to go to sleep. At last we dozed off.

We groaned with aching shoulders when we first got up at seven. Carrying heavy loads with our arms was doing it the hard way, we were convinced. We felt like taking it easy today! We wanted time for contemplation, to think about where we were and why. We needed to bask in the realization that we were out of the human rat race of civilization — to get the feel of the wilderness. During four previous summer wilderness experiences, we had found that it usually takes about three weeks to unwind from the pressures and tensions of civilization.

Gil built a small firepit and lined it with stones, then made pancakes for lunch. Yum! I devoured six. The day was spent loafing around, sitting in the sun and exploring the immediate area. A small spring of clear cold water bubbled out of the bank below our camp. Gil tied a line on a small bucket so we could haul up water, just like a bucket well.

On the riverbank, sitting on a log, we started getting acquainted with some of the wildlife. A friendly female pine grosbeak came close to us when we imitated her three-note whistle. We were pleased to read in our Peterson's field guide that they remain in the North year-round. A young northern waterthrush bobbed and foraged along the spring pool below. Several yellow-rumped myrtle warblers and a dark-eyed slate-colored junco flitted into the bushes. Then the junco pecked around on the ground at our feet.

With civilization miles behind us, we were apprehensive as we organized our tent camp. This would be our home until our cabin was built.

When it started to rain we built a larger makeshift tent with poles and tarps. A plastic window gave us light and limited visibility. Food and cooking supplies were arranged inside. We constructed a crude work table of poles outside, our "pole table." We felt we had all the amenities. Gil was to be the chef of our expedition and would do most of the cooking outdoors, but here was a place to eat out of the rain. I would be on clean-up and K.P. detail and keep our supplies organized.

"Before we go to bed we'd better have some kind of a bear-warning system. If any bears wander by we'd better know it." We made a barricade of poles, arranged in crisscross fashion across the bear trail, with an obstacle course of five-gallon cans stacked in front. Carabiners, large metal snap rings used by mountain climbers, hung on a line, to clang like a bell against our big handsaw.

"These contraptions won't stop any bear who wants to come through," Gil admitted, "but he'll make a lot of noise to wake us up." The trail ran within 10 feet of our tent. Any bear approaching from the opposite direction would hit another "Rube Goldberg" device. A nylon line across the path would pull and ring a bell, made of a skillet and a metal pan holder. We finished with a pole fence all around camp, like a corral.

"The trouble is if a bear sets off the alarm system he's already too close." Under those circumstances, I didn't expect to sleep soundly.

We were pretty jittery as we prepared to retire for the night. Gil's place was near the entrance of the tent. He placed the rifle on the floor beside him.

When we go into the wilderness we always bring a paperback library. We like to read aloud to each other, taking turns. Thus we can share the contents of the stories for discussions later. It's such a nice way to end the day.

"I'll read first tonight," Gil offered. "Let's read Crisler's *Arctic Wild* again." I listened, enjoying the story. A half-hour later I was getting drowsy.

"R-r-r-woof!" Suddenly I was wide awake. The woofing seemed to come from somewhere in front of the tent, down on the river bar.

Gil continued reading. Perhaps I had imagined it. No longer sleepy, I listened intently.

Then I heard it again, closer. My heart stopped.

"Did you hear something?" Gil asked.

"I sure did!" I sat up. "Is it a bear?"

"Could be."

We stood in the half-light outside the tent. The moon was just rising. Gil held the rifle, listening. The sound seemed to come from the willow patch south of camp.

"R-r-r-r woof." Some large animal was flailing the tall willows. Was it two animals fighting?

"Maybe it's a bear," Gil wondered. "No, it's a bull moose. I can see his 'boards' shining in the moonlight, above the willows."

Gil was chef and did the cooking outdoors.

With binoculars, I saw the broad, light gray antlers shining on either side of a tall willow. The grunting continued. Then the bull was quiet. I wasn't scared now. I watched awhile and went back to bed.

"He's in rut, that's what makes him act like that," Gil said.

The bull kept moving closer, still grunt-woofing, closer and closer. We heard brush cracking. Then he was getting too close. I got out of bed again. Was he getting ready to charge? He might be slightly demented in his present condition. Gil had his rifle ready for action. The moose crashed out of the willows right below camp, then swerved and raced toward the big river. We didn't hear him any more.

We went back to bed, but it took me a long time to relax enough to sleep. I stared at our flimsy canvas walls.

It was 9:00 A.M. when Gil got up to build a cooking fire, while I stayed in bed to write. I heard wings going "Sssh, ssh, ssh." A raven was flying over camp. He croaked. I answered him from inside the tent. He croaked again and went on his way.

Gil was cooking rolled oats for breakfast. I heard the sound first, like a low-pitched coyote howl.

"Listen. It's a wolf howling." I scrambled out of the tent. We stood spellbound, listening to throaty moans and howls from perhaps a mile upriver. What a thrill! The howls were the first wolf voices we had ever heard.

"Sounds like three of them," Gil surmised. Their clamorous voices resounded through the valley for several minutes. Then a low moaning reply came from the gravel bars south of camp. A raven circled over the willows and we knew where the creature was, although the shrubs screened our view. Twice more the beast howled, then silence. The circling raven told us where the wolf was moving. We heard a crackling in the bushes as it passed. Then we went back to our breakfast.

Fall trees glowed with reds and yellows.

Viv admired the sunshine yellow of willow and cottonwood on the river bars.

2·Building the Cabin

We chose a sunny spot at the south edge of the trees for our cabin. With the long dark winter approaching, we wanted as much light and sunshine as possible.

We began clearing the site. Time was of the essence. Winter comes early in the Arctic. With a hatchet, I whacked away at the brush, while Gil used an ax. A fox sparrow scratched energetically in fallen leaves under bushes behind camp.

"We have an audience," Gil alerted me. I looked up to see a young goshawk in a spruce about 60 feet away. A stubby-tailed raven watched from another treetop, curious and unafraid. It was pleasant to have the young birds around while we worked.

At this latitude it takes hundreds of years for a tree to grow large enough for cabin logs. For us it was unthinkable to cut live spruce. We would use only a few dead standing trees. Most of the logs would be windfalls which we had found lying on the ground or across other logs.

Near the site where our cabin would stand, Gil swung the double-bitted ax and felled the first tree.

"It's a shame to cut that tree even though it is dead. It's been standing there since before Columbus discovered America." I trimmed branches from another tree, one broken off in a storm.

During the past year Gil had read extensively on how to build a log cabin. He chopped the first notch.

"Beautiful," he exulted, proud of his handiwork. "This is easy." He finished rounding out the notch, rolled the log over, and it fit nicely on the log already in position on the ground. But it was beginner's luck. At times it didn't go so well. There were times when the notches were not in line, or there was a knot at the wrong place.

"It's past noon," Gil announced, glancing up at the sun. "Let's have lunch." His watch was packed away for the "duration." Who needs a watch in the wilderness? He was mixing up some pancake batter when we heard a new sound. He looked up from his cooking and listened intently.

Gil chops a notch.
He thought it was easy at first.
At times it didn't go so well.

17

*The rose hips were huge and
tasted like fresh apple.*

"It's a plane. No, it sounds like music, like a
drawn-out note on a pipe organ."

"It's wolves howling again." Gil turned from his
cooking. "Yes, it does sound like organ music."
The wolf chorus continued its lamentation, waver-
ing up and down the scale, for two or three
minutes. That was all.

The young waterthrush still bobbed and
teetered, while searching for food near the spring
pool below the cabin site.

With his streaked breast he looks like a thrush, I
thought, but he acts more like a spotted sandpiper.

Our cabin was to measure 12 by 14 feet inside.
We planned to chink the spaces between the logs
with sphagnum moss, which we gathered from the
ground under the trees. Because the lush soft moss
was a necessary part of the ecosystem, we were
careful not to denude the forest floor. Many little
creatures made their homes there, so we carefully
plucked clumps of moss here and there, spacing
them. The resilient yellow-green vegetation helped
retain moisture and provided insulation for the
roots of the trees and other plants, which included
a beautiful pink lousewort, blooming sparsely in
the moss, and red bearberry, which was in the full
glory of brilliant fall color. Overripe berries had
fallen from the plants.

"Those rose hips are huge. Hey, they're really
good to chew. They taste like fresh apple." We
nibbled on the tasty fruits as we carried large
bundles of sphagnum, rolled up in tarps, a quarter-
mile to our building project.

While Gil worked on another notch I placed a

*Red bearberry brightened the
forest floor.*

layer of moss in preparation for the next log. We
worked industriously, racing ahead of winter,
taking little time to rest. After supper, we often
strolled out to the river for a while. Finally, dog-
tired, we rigged up our bear-warning apparatus
and went to bed to read. Gil was now reading
Farley Mowatt's *Never Cry Wolf* again.

In the wee hours of the night Gil nudged me.
"Listen." I sat up half-awake.

"Yee-oww-oww-oww-oww."

It was a wild wail, lingering into a long, mournful moan.

"That's a cat," Gil muttered under his breath.

Again we heard the wail beginning like a high-
pitched siren and going down into a throaty moan.

"Yes. It's like a house cat's voice, but louder and deeper." It came from the underbrush behind camp. "It's a lynx."

"Pretty scary, if we didn't know what it was. It really sounds wild."

We heard the cat's complaining voice again, moving away.

Because Gil's watch was put away we'd get up in the morning when we were rested. I suppose that was usually about six or seven o'clock. It didn't matter. We weren't going anywhere anyway. We were always eager to begin work on the cabin immediately after breakfast.

"There are those strange bird cries again." We had heard these loud cries off and on for several days. "Wonder what they could be. They're in the woods east of camp. Let's go find them," Gil urged.

"Keee-keee-keee-keee." The call was loud and shrill.

"Pee-yop peeyop-peeyawp-peeyawp."

"Now, that sounds like a turkey," I said. "I know turkey calls."

"That one is rather like an osprey," Gil offered.

We hiked east of camp toward the calls. We found an enchanting forest of tall spruce trees. Our feet sank deep into soft carpeting of luxuriant yellow-green sphagnum moss, with great mounds of bright green velvet. Later we referred to this area as the Enchanted Forest.

The bird cries moved farther away, fading into the distance. Frustrated we returned to camp. The following morning the mystery was solved when two goshawks came into camp with the same cries.

We'd had one of the young hawks visit camp before, but it was silent then.

The young hawks were very noisy now, calling most of the time. They flew from one perch to another crying. We noticed their unusually large eyes, streaked breasts and fluffy undertail coverts. Gil's camera was always at the ready. What fun he had photographing these large hawks. They were totally unafraid.

We found some of the logs we needed several hundred yards away. They were large, straight windfalls, lying off the ground across other logs. First we cleared and smoothed a path for the moving project, then we used the primitive system

Viv placed a layer of sphagnum moss in preparation for the next log.

A tame young goshawk visited our camp. The young hawks were very noisy.

of pushing the logs on rollers we had cut from smaller logs down the path to the cabin site.

"It's amazing how easy it is to roll heavy logs by this method." I was surprised at how easily Gil maneuvered them.

"This is Swede logging," Gil grinned, teasing me about my Scandinavian ancestry. "You've got to be a Swede — or married to one — to work this hard."

"It's good for us to keep in shape. Whoa! You're going too fast." I was out of breath.

Gil pushed the logs down the Swede Trail, while I ran back and forth, placing the rollers in position ahead and retrieving them from the rear. It was a

great satisfaction to solve problems we had never before encountered. We felt a healthy fatigue after our day's work.

Nearly every evening we walked down to the river to look for wildlife and study tracks in the sand. But we didn't stay long. Our food supplies and gear could be demolished in short order if a grizzly wandered down the bear trail. Just the thought of it hurried us back to camp to be sure all was well there.

When we went to bed at dusk all was quiet and serene. Suddenly, out of the woods came the notes of a varied thrush, sweet and clear, like taps. "Good night. All is well."

Near right: *Gil pushed the logs down the Swede Trail. Viv ran back and forth placing the rollers in position.*

Far right: *The full moon came into view over the mountains.*

*Viv used a hatchet and chisel
to smooth out a notch.*

they were placed into position we stepped back to admire our work.

We were building our cabin on land administered by the Bureau of Land Management. The bureau had some problems with land investors trying to come up and homestead or to stake out mineral claims as a ruse for acquiring the land for real estate investment.

Carl Johnson, district manager, sensitive to preserving this "National Interest" land, helped us get a $10 permit to allow us to build the cabin, but we were required to sign an agreement to dismantle it before we left the site.

Carl dropped by, via small plane, for us to sign the final papers. He reminded us we would be subject to a substantial fine should we not abide by the agreement to dismantle the cabin.

•

In August the temperature dipped and vivid fall colors glowed everywhere. Fiery red dwarf birch brightened the slopes. River bars were clothed in the sunshine yellow of willow and poplar leaves. Cold drizzle did not dampen our spirits; doggedly, we hurried with our construction project. One thousand feet higher on the mountains, it was snowing.

Birds from the North Slope were hurrying through on their southward migration. On a lunch stop dunlins rested and probed the mud, their long sharp bills working like sewing machine needles. Bellies smudged with black made them look as though they had squatted in tar.

After one flock left, chirping like crickets, we

Proudly we watched our cabin grow. We found contentment in our work, finding the proper logs and trimming them, horsing them into position, rolling them down our Swede Trail. Gil put a lot of sweat into chopping notches, while I used a hammer and small chisel to smooth out a few. I was in charge of chinking between the logs. After

spotted a single dunlin squatting motionless at the edge of the shallow water. It looked much like another rock, half in, half out of the water.

"What's the matter with that bird?" Gil wondered. "Is it sick? It doesn't flush when we approach it." He glanced up. "Oh-oh. See that peregrine falcon perched right above it, in the cottonwood."

The shore bird didn't bat an eyelid.

Interfering with nature, I flushed the predator. When it flew away, the little dunlin stood up, shook himself, preened and nonchalantly began probing the sand again. The lone dunlin hung around for several days, foraging on the gravel bars near camp. We enjoyed its company.

Snickaree, one of the red squirrels, provided entertainment while we worked. Frenetically gathering cones and seeds, the accomplished aerialist skipped and practically flew through the trees "with the greatest of ease." It didn't take much to get her excited.

One afternoon we went down the logging trail to roll in another log. Before lunch the squirrel rained cones and seeds down on us. Now, on a log we wanted to move, she started to chatter, which at first sounded like a fast series of sneezes, followed by rapid-fire hiccups, which changed to ratchety coughs, then little barks. We didn't blame her for being upset. With a flick of her tail she ran up a slanting log which led up toward her nest. Then she ran down the tree and scampered back up a spruce on the other side of us, peering at us from beneath thick branches, where she berated

us again . . . at length. Then she skipped away through the underbrush. About that time, we heard the now familiar cries of the young goshawks coming our way through the tall spruce trees. We saw a dark adult coming, flying below treetop level, followed by its two progeny. The adult carried something in its talons, swung around sharply over the young ones and dropped the prey, which was promptly snatched in the air by one of the youngsters. "Keee-keee-keee," cried the young hawk who had received no food.

Gil and I returned to cutting and rolling logs. To utilize ends from logs we had cut we placed and pegged them in position, leaving openings for two windows on the south side of our structure. With a hand drill bit, I bored one-and-a-half-inch holes while Gil whittled long pegs to drive through into the logs beneath. This procedure deviated from the customary method of sawing openings for windows and doors after full-length logs are in place. We learned the hard way that when pegs are slightly too large in diameter they may stick and refuse to be driven into the lower log. In such a case, we would start all over again, drilling another hole and using another peg. Twice as much work. Later we learned we should have used square pegs.

Furry little creatures came out of the woods and shyly gazed at us before scurrying back into the forest. But one, a mouse-sized redback vole, became quite tame. Gil tempted it with an offering of pancakes. His long thick reddish coat glistened, his whiskers wriggled in excitement when he

Gil takes a break, happy to reach window height.

caught the enticing aroma of Gil's cooking. We invited him to dine with us, enjoying company for dinner. But we were careful to keep the rest of our edibles tightly covered in tins. His forest home was a burrow, hidden deep under the sphagnum moss carpeting the forest. We named him Eric the Red. A few days later there were two voles fighting over the crumbs we tossed them.

A family of wolves lived in the area. Frequently we saw both parents lounging on the gravel bars south of camp while we caught only brief glimpses of four gangling pups kept hidden in the thick willows behind them. The pups could be heard making small yelps and barks as they played and scuffled in the brush. The parents looked much like German shepherd dogs. They kept close watch over their rowdy youngsters.

For the first time in our several trips into the Alaskan wilderness Gil had hopes for some good wolf family pictures. These wolves had not been harassed by man. Fascinated, we watched them pass by, sometimes stopping to watch us, then going about their own business, while we concentrated on building our home.

One evening Gil took his fishing pole and tried fishing in the river. He caught one fish, anchovy size, and tried again.

"Only fingerlings seem to be here. I wonder where the rest of them are." Back at camp, we settled for a cheese omelet for supper, made with powdered eggs and cheddar. It tasted good. We had good appetites.

Late in the evening as the sun was about to go

down in the northwest, a lone wolf howled from across the river. Then a number of young wolf voices joined in.

"Those sound like the half-grown puppies," Gil noted. "Several of them."

The wild voices came from downriver. They were mournful cries.

"You know what that sounds like? Like a domestic puppy cries when taken away from his mother. Sounds very sad."

Gil got up on the unfinished cabin walls and looked south.

"I see a wolf. It has an injured right front foot. It's more brownish than the pair that we have been seeing around here. It's moving this way, going right. Let's run out to the bar in that direction. Maybe we can get a picture. Let's be quiet."

With cameras and rifle, we hurried down the bear trail, down the bar and sneaked up behind some willows. I stayed behind Gil. We didn't have long to wait.

"Get down. I see it."

We watched the hurt wolf come limping. It was gray with brownish tips on the ears. It scented us and stopped. Then it went down to the edge of the river. A pup arrived, of the same color, about like a three- or four-month-old German shepherd, with big feet. The adult went into the river and started to swim across. The youngster got ready to follow, hesitated a moment, then was in the water, paddling across against the swift current. Another pup was following. It had appeared about 40 feet behind the first. It began to swim across. Two more

A redback vole became quite tame. Gil offered him some pancake.

pups appeared. All were the same typical German shepherd gray color with reddish ear tips.

Breathlessly we watched the wolves proceed in single file; while the adult wolf was shaking itself on the opposite shore, two young were swimming across, another entering, and the fourth approaching ready to enter the water. The limping adult led the way into the willows. The young wolves followed in its tracks, all keeping about 40 or 50 feet apart, as if they had been trained for it, even while swimming across, backs just visible, paddling hard to keep from being swept downstream. When they reached the opposite shore each stopped to shake water from his fur, then trotted on in the tracks of the leading adult. At last the fourth pup awkwardly shook himself hard and was soon swallowed by the shrubbery. We were spellbound to have witnessed this river crossing by the wolf family.

Were they on their way to join the wolf who had howled first? Was it their father? I visualized the rambunctious pups joyfully greeting their parent. A little later we heard puppy howls from across the river. Walking back to camp, we heard the pups howl again.

"There's a lynx." Gil pointed downstream from the spring pool. With binoculars, we watched a light tan cat stalking, treading softly with its large paws along the edge of the open bar. It turned into hip-high willows, its tawny rear accented by the black tip on its stubby tail, disappearing into the brush. We waited for another view of it, but were disappointed. We hurried down to the bar, hoping

to catch another glimpse. Suddenly the wolves howled, long and loud, from across the river. An answering howl came from our side of the stream, not a wolf howl, but the wailing moan of a cat. It was the lynx in the brush about 100 yards away. Then a cacophony of wild cries began, mournful, lonesome wails from the cat, like one who has lost his last friend, on our side of the river. The wild howling of the wolves replied from the opposite side.

"Like mourners at a wailing wall, trying to outdo each other."

In the half-light, we stood in waist-high willows, listening to the wild laments of the untamed. I felt a prickly sensation on the back of my neck.

"Now all we need is a grizzly to show up." Gil pointed to taller willows 50 feet away, which screened whatever was there.

A shore bird flew over our heads, upstream, then wheeled and came past us again. It settled down in the yellow dryad which formed a thick carpet on the bar. It was a pectoral sandpiper.

The cat let out three more agonized moans. A short howl came from a young wolf, then silence. We stood waiting, listening, holding our breaths, looking at the deep crimson color in the northern sky. Reverently we walked back to camp, grateful for this precious evening with its wild concert. Down the bear trail, we stepped delicately over the bear barricade to our unfinished cabin and tents.

As the sturdy log walls grew higher, we took comfort from our partial fortress, should a grizzly amble past. The flimsy walls of our sleeping tent

The tracks of an injured wolf showed drag marks.

We hoisted and placed three logs for ridgepoles. This called for a celebration.

offered protection from wind and rain, but wild cries sometimes jolted us wide awake and made us light sleepers.

From my diary: *August 12. Today I spotted Granddaddy Bull Moose near the top of Noon Mountain, directly to the south. Sun was shining there in late p.m. With magnificent rack, he moved regally through short willows toward the left. Gil has plans to get him later for our winter meat supply. I abhor the thought, but it may be necessary. I put it out of my mind as we begin to study animal tracks.*

We took a walk by the river and Gil took pictures of tracks with drag marks made by the crippled wolf.

"This means the leg is injured, not the foot," Gil explained. "Perhaps it's broken."

•

Weeks flew by. When the side walls were five feet high, we began pegging shorter logs into place for the gable ends. Gil used the ax to trim them to the proper angle. Triumphantly we hoisted and placed three logs for ridgepoles. This called for a celebration. I took Gil's picture on the ridge and he served a special dinner of delicious smoked salmon (a gift from his student, Jon Walden) with steamed rice and finished off with cheesecake pie.

We were anxious to get a roof on our dwelling without wasting time. We searched the woods for thin straight spruce poles. Spruce is relatively lightweight and strong. A half-mile away, near the Enchanted Forest, we found a thick stand of small,

dead spruce. We cut and carried them to camp, three or four at a time. Each had to be trimmed of branches and cleaned of bark before being placed for our roof. Gil tested each one by hanging with his full weight on it, before it won his approval. Then he cut them to the proper length and handed them up to me on the roof. One hundred forty small poles were placed side by side to form a solid base for the sod. Not a nail was used. The weight of the sod would hold them in position securely.

Our cabin already provided shelter for us. When it rained we covered the roof poles with plastic sheeting until the shower ended. We gathered more sphagnum, boxes of it, and placed a four-inch-deep mat of moss over the poles.

Gil shows off the 30-gallon drum which was converted to our stove.

The opening for the stovepipe was through a square five-gallon can fastened to the roof poles. First a sheet of plastic, then a layer of sod was added over the moss. For more headroom inside the cabin, Gil dug the sod out of the floor and handed the dirt up to me. The roof was completed just five weeks after we started construction. None too soon!

Fall rains began in earnest. Happily we carried all our possessions inside our new house. Feeling smug, we stood in the doorway and watched the rain come down in torrents.

"It's a little chilly in here. Time to set up our stove." Gil brought in the 30-gallon oil drum he had converted into a barrel stove at Bettles. We stacked some rocks in the middle of the room to support it, then attached a stovepipe oven, a generous gift from our Eskimo friend, Frank Tobuk, of Bettles. Stovepipes were run through the five-gallon can in the roof. Soon the fire was blazing and we felt blessed warmth fill the cabin. We admired the stove, then went out to exult over smoke coming out of the chimney. What a happy sight! Smoke meant warmth and cheer and survival through the arctic winter.

We covered the roofpoles with sphagnum. Over this went sheet plastic and a layer of sod.

3·Death of a Wolf

 One frosty morning in early September we heard a whistle outside our cabin. Two men walked up the bear trail. Pat Yaska, an Indian, and Jim Edwards, an Eskimo, from Allakaket, had traveled upriver in a longboat for some hunting. We invited them in for tea.

They were friendly fellows, and told of an old structure, now tumbled down, which was across the river. In the old days it was used by the Natives for drying and smoking meat and fish.

"Eskimos used to come up here by boat and camp," Jim said. "They would hike up North Creek to hunt sheep, then pack out the meat in relays. The meat was smoked in a shed. About 20 men would spend all summer to hunt and fish up here, then return to Allakaket with meat in the fall."

Pat and Jim packed up their gear for a hike up North Creek to look for sheep. Pat carried a caribou skin for a sleeping bag, while Jim had a contemporary down bag. They planned to return the following day.

The next morning Gil and I saw the smoke from their campfire and visited them on the shore of the big river. They had seen no sheep. But that didn't dampen their spirits, even though it was raining. They were comfortable in a tarp lean-to, which reflected heat into their shelter from a cheerful campfire in front.

Gil invited them for breakfast, so they came to our cabin to feast on Gil's pancakes. Later we bade our new friends good-by as they headed downriver to hunt for sheep, caribou, moose and bear. We wished them good luck.

Gil planned to go moose hunting, too. In our wilderness food supply we included some tinned meat, fish, cheese and powdered eggs, but we needed more protein than that.

Before we left civilization, Gil bought a nonresident hunting license and a moose tag. He's a crack shot with a rifle. But he has made eye contact with too many wild creatures while photographing them to find joy in killing.

"If I get a big moose, we will have enough meat for a year, and I won't need to kill any more

*Our cabin was reflected in
the pool from a spring.*

mammals. Some fish caught now and then would be good for a change."

We began building a meat cache in tall spruce near the cabin. We lashed a pole platform with nylon lines 14 feet from the ground and built a pole railing around it. By using lines it could be disassembled easily when we were finished with it.

The cache was not quite finished when Gil decided it was time to hunt for moose. At the crack of dawn on September 18, dreading the distasteful task, he went out to the bar and shot a big bull moose, forcing himself to pull the trigger.

When I heard two shots, I ran out and found him on our airstrip. I cried when I saw that it was Mighty Moose, a splendid animal we had been observing for weeks, lying in a grotesque position on the sand next to the river.

"I just about let him go," Gil grimaced. "But we can't survive without meat. He gave his life for us."

"It was unfair. He didn't have a chance. He probably had as much right to live as we do."

"What bothers me is remembering how I used to hunt for the sport of it." Before taking up wildlife photography, Gil went hunting for deer and elk in Oregon every fall.

I remember that barbaric scene. I was crying and Gil was none too happy as we skinned, eviscerated and butchered a beast which seemed as large as a mastodon. While we were thus engrossed, a young bull moose wandered by through the willows. We continued our gruesome task.

"Hey, that young moose is coming back." About 40 feet away, the young beast stood watching us at our bloody work. My feelings of guilt were intensified.

A couple of gray jays flew up to investigate. They spread the word among their buddies and soon a dozen jays began cleaning up the scraps we tossed aside. Busily they carried bits and pieces in all directions, stashing them in their favorite hiding places. One jay was carrying a hunk of moose meat across the river. It was so large that he could hardly hold it in his beak.

"He'll never make it." Gil was amused. "He's

Viv saws a plank. It took two hours to make one cut.

losing elevation." But when the bird was down close to the water, we saw him transfer the meat from his beak and grab it with his claws. He proceeded, flying laboriously, just above water level, across to the other side of the stream, hanging onto his loot with his feet.

"I didn't know they had that much strength in their feet," Gil remarked. "It's easy to understand that they have the intelligence to solve a problem."

Becoming more confident that we were no threat to them, the jays continued collecting meat scraps around our feet within inches of us.

We heard a sudden whirr of wings. A peregrine falcon shot close overhead. The jays dived for cover; one cowered by my feet; another slipped in next to the carcass, hidden by brush. The falcon perched in a nearby small cottonwood for a few minutes. Not a jay moved a feather. Then the hawk flew away.

With quarters of moose meat swinging between us from a pole on our shoulders, Gil and I worked in relays, carrying a total of about a thousand pounds of meat and bone three-quarters of a mile to our cabin. Darkness was overtaking us when we got it home. We wondered wearily where to put the meat to keep it safe from bears.

"Let's sleep with it tonight." I was exhausted. "We can finish our cache tomorrow."

So we dragged the meat inside the cabin and put it on plastic sheeting on the dirt floor. It had been a strenuous day, both emotionally and physically. We slept soundly. What a comfort to know we had our year's supply of meat!

Gil shook me awake next morning. "We did a foolish thing. If a bear came by and smelled that meat he would walk right in and help himself." We had no door. Only a plastic curtain hung over the entrance.

My eyes popped open. I sat up.

"I'd hate to have a fight with a bear in here," he added.

So we finished our meat cache, with a sturdy pole placed horizontally about six feet above the platform to support a pulley system which we rigged to winch the heavy hunks of meat up onto our cache. Our 15-foot ladder had rungs lashed in place with light nylon line.

"Now we need a door to our cabin. That won't be so easy."

We got planks the hard way, hand-ripping them from six-inch-diameter logs. It took two strenuous hours to make one cut four feet long and six inches wide. Taking turns, we sawed four planks from each log. The curved planks for the exterior would be used vertically, while the flat planks would be used horizontally on the inside of the door. So far we had used no nails in our cabin construction. But we did have a couple of pounds of nails. We used most of them now to nail the planks together, with sheets of corrugated cardboard and plastic sheeting between the two layers of planks. When finished, the door was four feet high, two feet wide and about four inches thick. It must have weighed more than 40 pounds.

"We need strong hinges to support the weight of that door." Gil wondered how to manage that.

A log could be sawed into four planks eventually.

Gil carved massive hinges to hold our bulky door.

"Why don't you carve strap hinges out of spruce?" I asked. "Spruce is strong."

While he carved massive hinges I cut and fitted planks to frame the opening, then chinked the cracks with moss.

We nailed the long strap hinges, with one-and-a-half-inch holes bored at one end for the pin, to the inside of the door. Then, because the door frame would not support the weight of the door, we fastened the big wooden hinge pins to the log wall.

It was an exciting moment when Gil hoisted the bulky door and hung it on the hinge pins. Miraculously, it fit. But when he swung it open an agonizing shriek sent shivers up my spine.

"The Inner Sanctum!" We laughed.

"Yes. I'd forgotten about that old radio program."

"I'll fix it. All it needs is some lubricant." Gil brought out a candle and waxed the wooden hinge pins. Then the door worked like a charm, swinging smoothly and easily without a whisper of a sound.

"Hey, this is great! We have a genuine door."

"You know, the door is a marvelous invention." I opened and closed it several times. We admired it from outside and in. It was strong enough to hold back a bear.

To enter our house, we had to duck our heads down, while at the same time, we lifted our knees high to step over the elevated threshold, then dropped down to a lower level inside, sometimes stumbling on the threshold. Later we became expert at this and scuttled in like rabbits into a burrow. On exiting we had to double up even

It was an exciting moment when Gil hoisted the bulky door and hung it on the hinge pins.

more for a knee-high step, then twist to sidle out the narrow opening.

Our threshold soon acquired a name, the "stumbling block." It took a while to get used to it.

Then our wilderness home was complete. We had made it ourselves and it was beautiful. We loved every log, every bit of moss.

"It looks like it belongs in Ibsen's play, *Peer Gynt.*"

"Now that we have a strong house, we can take down our bear-warning apparatus." We dismantled the noisemakers and barricades across the bear trail.

Only one incident marred our happiness. It concerned the wolves. Shortly after our arrival in the wilderness, a prospector, with two young men,

came through the area. They camped across the river. Many Alaskans persecute wolves so we feared for the animals' safety. We hoped to keep the wild canine family a secret from the party across the river. Whenever the wolves howled we worried and hoped the men didn't hear them.

One day the wolves set up a duet from south of our camp. Standing atop our roof, we saw the parent wolves on the gravel bar about a half-mile away. We hiked toward them so they would run away. The adults sat on their haunches, like two big friendly dogs, tongues lolling out. We could hear the pups whining in the willow patch behind them. But they wouldn't spook; they just sat and watched our approach. They trusted us.

"Let's give them a good scare so they'll disappear for a while," Gil suggested. We ran toward them, yelling, screaming and waving our arms.

The big wolves fled, quartering, making a big detour around our cabin and headed through the trees for the river.

A short time later we heard two shots.

We ran to the river and watched one of the young men wade the stream to our side. He was carrying a rifle. He entered the woods, looking all around, obviously looking for a dead wolf. He found none. The prospectors left and we never saw them again.

From the moment of the shooting incident the wolves avoided us like the plague. A few hours later, mournful howls came from near the river.

"Gee, that wolf sounds sad, like he's crying," I told Gil.

"Yes," he agreed. We listened to the wild lament, starting on a rich, high note and sliding down the scale.

"It still sounds like organ music." I listened spellbound. "It's beautiful."

"Sure sounds wild."

Every day and night we heard the wolves howl and one cried out with a distinctive wail. We came to listen for him and named him The Old Mourner. The howls usually emanated from the same general area in the trees, near the river.

•

Now that our house was built, we settled down to enjoying wilderness living. Our year's meat supply was safely up on the cache. We gathered some windfall spruce poles and stood them on end

It was strong enough to hold back a bear.

near the cabin for an emergency wood supply. Even in deep snow they could always be found.

Around a bend in the river upstream, Gil caught some good-sized grayling and we had a feast, all the fried fish we could eat with instant mashed potatoes. There was some fish left over to store up on the cache.

Immersed in the distractions of living in civilization man becomes desensitized to nature. In the wilderness we became attuned to sights, scents and sounds, which ordinarily we would have ignored. We roamed, watching, listening, enjoying. We welcomed the animals, feathered or furry. They were our neighbors. We studied their personalities, keeping detailed records of our sightings and their behavior.

One bright fall day we were in high spirits as we started hiking upriver on the gravel bars. We had not gone far when Gil stopped.

"Something stinks! Do you smell it?"

"Sure. How could I miss it?"

"It's decaying flesh." He put down his pack. "I'm going to see what it is."

While he went into the trees on the riverbank I watched a couple of jays, who were watching me.

"Guess what I found." Gil called from the bank. "A dead wolf! There's a big hole in its side. It was shot." He sounded bitter. "Those fellows killed it."

I ran over and he showed me the carcass, partly covered with new fallen leaves.

"It's the mother of the pups," he informed me. "Isn't it from this area that we keep hearing that mournful howling?"

"Seems like it. Do you suppose it's her mate?"

"Could be. But we can't be sure. When the government 'wolfers' were exterminating wolves in the Lower 48, if they trapped a wolf, it wasn't too hard to get the mate. The mate usually came back looking for its lost partner."

"What a rotten business!" I felt depressed. "Now this! It's gross."

"There goes our only chance to observe, photograph and get acquainted with a wild wolf family." Gil was crestfallen. I knew how much this opportunity meant to him.

"Isn't it illegal to kill wolves in Alaska now? Isn't it out of season?"

"Yes," Gil answered softly.

We continued our outing up the river, but our day was spoiled. We could not get the wolf tragedy off our minds.

"I wish more people could have our experience, living with wildlife in the wilderness, studying animal behavior. When will they ever learn?"

"If they go planning to shoot they will never learn."

During our many months of studying wildlife unmolested by man, we learned much. We noticed each animal has a distinctive personality. Now we regard them as individuals with feelings.

People in civilization who keep pets become quite attached to them and consider them individuals with certain rights, like members of the human family. If they have not known wild animals it does not occur to them that these, too, have personalities.

We have more admiration for free-living wild creatures than for spoiled and pampered domesticated pets. In an ever-eroding habitat the wild ones must take care of themselves, pitting their intelligence against the arrogance and cruelty of man, and the encroachment of exploding populations of his pets. These pets seem to us to compete with man himself in a protein-short world.

During the nights of mid-September temperatures dipped below freezing and a skim of ice formed over the quiet pools. After a long period of rains we welcomed a few sunny days. Thinking of the coming winter darkness we grasped every bit of sunshine we could get. We were greedy for sun and sat on a bench on the south side of our cabin soaking it up.

Diary: September 18. *24° when Gil got up at dawn to cover the meat on the cache with netting before the jays get into it. They seem very greedy, but, when we realize they are stashing food for the long cold winter ahead, they are wise. It's what we are doing too.*

Gil catches more grayling. Rolling them in corn-meal, he cooks them in a skillet over the open fire.

That fried fish is the most delicious I have ever tasted. I go into ecstasy over the excellent flavor, as we sit on the bench and eat in the sunshine.

We watch spruce grouse stroll around our yard, tame as chickens, making little clucking sounds. They pecked around, rearranging fallen leaves on the ground, searching for food. A couple of snowshoe hares live in the vicinity, hopping leisurely around, nibbling on alder leaves and various other plants. Gil is trying to win their confidence by humming softly while photographing them. Eric, the redback vole, and a companion vole, dart around the yard at noon. We try to keep them out of our house. So far, we've been successful.

To furnish the interior of our cabin, we hand-ripped boards and built a counter-type table, 14 feet long, along the south wall. This served for food preparation, dining, writing and other purposes.

We made our bunk bed 39 inches wide and about 4 feet up from the earthen floor, with plenty of storage space beneath. Small poles lay close together to support us. No springs, of course, but for a mattress we had a large muslin bag, which we stuffed with sphagnum moss until it was 18 inches thick. Oh, it was deliciously soft — the first night.

"This mattress is lumpy," Gil complained the next morning. We arranged the moss and added more to fill the hollows as it began to pack down. For weeks it was lumpy. We kept smoothing the moss and adding more stuffing, while the mattress got firmer and firmer. We stopped trying to keep it soft, but concentrated on the lumps. At last we were content to have a fairly smooth hard bed. We got used to it.

"This mattress is getting so hard, it should test about 3 on the hardness scale, as geologists classify rocks," Gil commented as he rapped his knuckles on the mattress.

On September 21, we started out early to climb

Ram Peak to the northeast. We had named it for two sheep we saw resting on its high ledges. After being tied down to our construction project and guarding supplies for so long we felt free to explore then.

It was frosty in the shade, but warm in the sun, as we hiked in sweat shirts with no jackets up the bars of North Creek. Soon we took to the riverbank, hiking on a lovely wildlife trail through thick soft moss, under spruce and small paper birch.We headed for the mountain on a network of animal trails. A jay accompanied us. After filling our canteens with water from a small stream, we began steep climbing through the taiga, the evergreen forests of the subarctic. Ripe red bearberries hung like translucent rubies on bushes. We were enthralled by the beautiful terrain, the unspoiled wilderness — no evidence of man, so lovely, so perfect. The jay followed.

It was as warm as a summer day on the south slope of the peak. We looked down to our cabin grove to get a new perspective on the area. Our wilderness home was concealed by the trees, but how comforting it was to know that we had a cozy haven to return to. How wild the land was! The reality of our solitude overwhelmed us. How grateful we were that there were wild places still, where one could find the world as it has been for eons — and not despoiled by man.

We climbed up farther, feeling fit and at peace with the world. What a joy to be there in the warmth of the midday sun!

Gil and I sat on a high ridge to eat lunch and watched a golden eagle at play. We first noticed him in a gully below. He rode the updrafts, with pinions spread, circling gracefully, slowly, until he was higher than the mountain. We watched the great bird fold his wings close to his sides and plunge down, like an arrow, until he was far below. Just before he reached the valley floor he spread his wings slowly to brake his descent, then let the thermals lift him again over the mountain, as high as before, where he repeated his sky dive at the same place. Again and again the majestic bird ascended the invisible elevator. High in the air again, he was a free spirit, sometimes barrel-rolling or somersaulting for the sheer joy of it. His exuberance was contagious.

A view of our river valley. How wild the land was! The reality of our solitude overwhelmed us.

The eagle soared close above us, turning his head to look us over, then went back to playing again.

"It's always a joy to watch wild animals at play," I told Gil. "What are those white birds? They fly like pigeons."

Looking below we saw a flock of about 25 snow-white birds, flying with fast, even wingbeats — a close flock, flying as one unit, immaculate white wings flashed in the sun.

"Ptarmigan," Gil told me, "in winter plumage." They alighted on the slope below, and their scintillating wings were still.

Down in the valley we could see that the rivers had many channels; they were braided streams. We studied with binoculars to orient ourselves to the terrain and plan places to hike.

"See that straight line across that channel downstream?" Gil pointed. "That must be a beaver dam. It's not natural for a line to be that straight across a stream."

"The Eskimos said beaver don't come this far upstream."

"It would be interesting to go down there and have a look."

"Hey! Here's the jay. He followed us up the mountain." Gil tossed him a piece of pancake.

On the opposite side of the valley, beyond the rivers, jagged peaks crowned the mountain ridges, like a dragon's back.

"Those are the Arrigetch Peaks," Gil explained, studying a topographic map. "Only granite spires are that jagged. It's truly wild country. Carl Johnson, district manager for the Bureau of Land Management, said that when he hiked in there last year prospectors were dynamiting. It really disturbed the serenity."

"Seems sacrilegious," I muttered. "Why do they let them dynamite in there?"

"Don't you know the world is short of hard minerals?"

"Gee. And to think of all that is wasted." I thought of countless beer and soft drink cans littering roadsides all over our country.

Suddenly a raven swooshed close over our heads, following the crest of the ridge to the top of the peak, where he barrel-rolled and played with the wind. It was time to return to our wilderness home. Happily we dug in our heels, descending the steep slopes on animal trails to delightful taiga — stunted spruce, bearberry, kinnikinnick, Labrador tea, buffaloberry and scrubby willows.

Many wild creatures lived around our cabin and we enjoyed all of them. They were our neighbors and many became quite tame. As they skittered around Gil took their pictures while we observed their behavior.

Eric, the redback vole, was a charmer and one of my favorites. We also had a pet ermine. One day the ermine ate Eric.

Several days later we set off to see if there really was a beaver dam. Gil led me through a maze of channels downriver. Ice was forming along the banks and on the quiet waters. Streams were lowering with the coming of winter. We rounded a bend.

A wet furry head with a puppy face appeared in the water.

October 22. We looked to Ram Peak in a world frosted with snow.

"By golly, there is a beaver dam!" Gil said.

"Beautiful! How do they know how to build it?"

Those intrepid engineers had used willow and cottonwood to construct a dam about 30 feet long. The pool above the dam was partly frozen over.

"Look, there's their lodge on the bank."

A couple hundred feet up from the dam was the beavers' mud and stick fortress. Open water remained in front of their dome house.

"They really work to keep open water as long as possible. Let's be quiet and maybe we can watch them."

We sat on the bank opposite the lodge. Soon Gil nudged me.

"There's one. Hold still."

A wet furry head with a puppy face appeared in the water from behind some debris. Puppyface dived, reappeared, and started swimming our way.

"Don't move," Gil whispered. "He sees us."

His round leathery nose swam directly toward us until the ice blocked his way. He wriggled his nose, testing the air for our scent. Gil and I didn't move a muscle. The creature contemplated us for a few moments, then swam back to join another beaver who shared the pond with him. He swam alongside his pond-mate and we heard low murmurs as they touched noses and conversed. The two of us sat quietly on the bank watching the two beavers bring up small branches of poplar from the bottom of their pool. They held a stick in their hands to chew off the bark, like corn on the cob. When one branch was peeled, a dip to the bottom brought up another delectable one. We heard

a crunching sound when they bit off another chunk of bark.

"Their food supply is at the bottom of their pond." I spoke softly so they wouldn't be alarmed. "What would happen if it got so cold the pond froze solidly to the bottom?"

"They wouldn't make it," Gil told me. "That's why there are few beaver here."

Winter was setting in. On the first day of October we had our first snow. It was time to install the eight small glass window panes Gil had brought in at the bottom of his camera bag. The plastic over the windows was replaced as fast as Gil could carve supporting frames for the glass from one-and-a-half-inch-diameter willow poles, anchored vertically and grooved to hold the panes. Slats were tacked on the logs to fasten the top and bottom edges of the glass.

The clouds lowered and a silence hung over our world as the streams got lower and lower, smaller and smaller. Snow muffled sounds. The sun dipped on the southern horizon.

Every day we went on hikes in the wintry scene. The gravel bars were frozen now. Icy boulders protruded from the snow. I noticed some frozen stalks of arctic fireweed.

"I'm hungry for fresh greens." I sampled some of the frozen fireweed. "Not very tasty." I spit out a mouthful.

"Wonder when the caribou will be migrating through from the North Slope?" Gil brought up this subject every day. This exciting event had been uppermost in his mind for several weeks.

Gil carved frames for the glass in our cabin windows.

4·The Great Migration

October 11. *There was a grunting snort coming from the direction of the big river.*

"That's a moose — or maybe, a bear." Grabbing jackets and binoculars we ran through the trees to the bars. Something rattled the river gravel upstream. Caribou!

Holding very still, we watched them approach. A cow was leading a single-file procession. We positioned ourselves at the edge of a patch of willows and Gil got out his camera. The deer kept to the middle of the open bars, marching toward us. We held our breaths. Would they turn before they reached us? The wind was in our favor. Crossing North Creek at its mouth, they seemed undecided upon which way to turn, then proceeded our way. One cow started to head east of us. Would she spook and rout the whole herd? But they came on, within 100 yards of us, passing us, not noticing. Gil snapped the shutter. Two or three bolted for a few yards, then turned and stared back at us. A cow led the line. About four cows, with three calves, then a splendid bull, another bull, more cows and calves was the order of the procession. Standing on the bar about 70 to 100 yards from us the bulls snorted and jousted briefly.

"There are 35 caribou altogether." I had been busy counting. A cow turned and led the band back across the big river, splashing four or five abreast, in the knee-deep stream, then shaking themselves, water spraying out, on the opposite shore. The white on the underside of their tails waved like so many white flags. They trotted into the willows and were gone.

We waited for more of the caribou to appear. After a half-hour we were chilled and returned to the cabin. Later we went back out to watch and wait. No more came. We went back to the cabin for a quick lunch while we kept listening and going out to check.

In the afternoon Gil said he was going out to wait for more caribou. "I'll wait until the main herd comes," he said.

Taking his cameras, about ten rolls of film, and a paperback book, *A Naturalist in Alaska* by Adolph

Caribou marched past in their luxuriant fall coats. They appeared fat and healthy. We started counting.

Murie, in his Kelty pack he went out to the same spot at the edge of the willows and settled down to read.

I stayed in the cabin to finish kitchen chores. Later, I heard a rattle and the snap of a branch. It seemed to come from the bear trail.

"Gil is making a lot of noise going through the brush," I thought. "He'll never get close to caribou that way."

Then I heard more. Could it be a bear going through the woods? Taking binoculars and a jacket, I went out to investigate the sounds, moving cautiously.

Something was rattling on the gravel. Sneaking closer, I looked through the trees. White fur capes were moving on the river bars on the other side of the trees in an unbroken single file procession. I watched the parade of caribou filing past, bulls snorting now and then. Their hooves rattled on the rocks. I heard them breaking the ice farther ahead. I wanted to share all this with Gil. I should alert him that the deer had arrived.

Turning, I ran through the woods, cutting across the willow patch and breathlessly joined my husband by the river. He smiled and pointed to a spot near the airstrip. I was thunderstruck. Countless gray and white furry beasts were milling around on the gravel bars on both sides of the river. Antlers of all forms and sizes protruded in a maze from the milling mass of deer. Meanwhile many animals were filing past, close to us, within 50 feet. A curious clicking sound came from their hooves.

Gil was snapping pictures like mad, his frantic efforts evident from the crumpled film wrappers strewn around his pack. He was operating in 35mm color and 2¼ x 2¼ in black and white and two kinds of color.

Barren Ground caribou surrounded us. They marched by, sometimes bolting and running a few yards, then turning to look intently at us for a while before they relaxed and paid us no heed.

"So many calves," Gil mentioned. I noticed it too. Nearly every cow had a calf, which looked fat

The bull was magnificent with his lordly antlers and flowing white cape.

and almost fluffy in his new light-colored winter coat. Some of the youngsters had nubbins of antlers, while others had spikes.

"A cow was leading this enormous herd," Gil told me. "Some fuzzy tatters of velvet were hanging from her antlers. Her calf followed closely."

Bulls, magnificent in their lordly antlers and flowing white capes, were grunting, snorting and puffing like steam engines. Each bull had from three to seven cows ahead of him. They were his harem, no doubt. A couple of smaller bulls tagged along. Occasionally, the big bull would turn and confront his male followers, asserting his dominance by grunting, puffing and snorting. When challenged the bulls lowered their heads and clashed their antlers from side to side. The cows could care less. They plodded along, closely followed by their calves, or lay down to rest briefly, chewing their cuds, gazing around as though bored with male chauvinism.

As the caribou marched past in their new luxuriant fall coats, they appeared fat and healthy. I began counting, "Fifty. One hundred. One hundred and fifty. Two hundred!"

"How many do you estimate are milling around here?" Gil was still busy with his cameras.

My eyes swept over the milling mass. Some stood across the river, some on our side. Caribou waded across the river. Caribou still filed past us.

"I would say at least 1,500, maybe more."

"That's just what I figured."

And they were still coming.

It was like a dream. They marched in, as calmly as people arriving for a convention, but quietly. No loud voices, except for a low grunt now and then. They bunched up and milled around. Cows were placid, bulls unsettled and restless. There were a few hassles. Calves, tagging their mothers, had no time for play.

"It seems the bulls' cumbersome antlers would be a handicap. The smaller antlers of the does make more sense against predators," I thought out loud.

"Must not be," my naturalist assumed. "That's

A cow, a calf and a bull filed past close to us.

the way they evolved. The strongest assert their dominance for more cows in their harems."

We were at peace with them. Some lay down to rest. The entire immense congregation stayed for more than an hour, over two thousand wild hearts beating, aside from ours.

The huge assemblage of caribou split into three groups. Several hundred crossed the river and were moving up the lower slopes of Noon Peak, the white patches on the caribou blending with white patches of snow. As they began grazing among the scattered spruce, I saw two calves romping in the snow, galloping in circles back and forth, holding their tails high, kicking up their heels. The herd fed as they moved slowly up and over the mountain.

The rest of the caribou headed downriver. We followed for a while and saw that a good-sized herd split off from the main herd again and climbed up the mountains to the left. The remainder continued down the river bars.

Walking back toward our cabin, we noticed that the gravel bars looked like a corral chewed up with hoof prints.

"Here come the stragglers." A few caribou continued to trickle through.

"Where will these animals spend the winter?" I asked Gil, an avid reader of everything about arctic wildlife he could lay his hands on. A number of scientists had studied caribou migrations.

"I guess they break up into bands and scatter as they move south, but I'm not sure if much is known about where they go. It seems to me that the darkness of winter would prevent close observation."

A few days later found us heading up North Creek. We met a couple of small bands of caribou. They spooked and ran into the woods. They were skittish and didn't have the confidence of the massive herd — totally different.

We climbed a ridge and scanned the slopes using binoculars. We discovered caribou in small groups, high on the mountains, above the trees, feeding on the short vegetation of the tundra. They seemed settled, as if they planned to stay.

Wolf tracks mingled with caribou tracks on the gravel bars. "Wolves belong with the caribou," Gil commented. "They cull out the sick and the weak."

The next morning Gil and I were building a table in the cabin. Just outside the door, Gil was whittling a peg for the table.

"Honey, here's a wolf." His voice was low and urgent. He didn't move a muscle.

I looked out and saw a beautiful wolf about 50 feet away, under the cache. Like an Alaskan malamute dog, it was silver, black and white, with thick fur and a heavy gray ruff.

The wolf gave Gil a casual glance, looked up at the cache, tested the air, sniffed at the moose antler, then looked up at the meat again, probably wondering how a moose carcass got up in a tree. The wolf circled the cache a couple of times, glanced up once more, then casually walked away.

"Darn! I didn't dare move to get my camera." Gil straightened up from his crouched position. "I was afraid I'd break the spell."

Wolf tracks mingled with caribou tracks on the gravel bars. Wolves cull out the weak caribou.

The huge assemblage of caribou split into three groups. Several hundred crossed the river.

Later Gil and I checked the tracks on North Creek and found there had been two wolves following the caribou trails. They encountered our tracks and followed them right to our yard.

It was a red-letter day for us. With pulses racing, we had shared a few moments with a beautiful, wild wolf, who was curious and unafraid.

•

We watched with mixed feelings of apprehension and wonder as the sun made a lower arc to the south. When the blazing orb touched the mountain at midday and disappeared for a while we knew it wouldn't be with us much longer. What would it be like when the sun was gone?

Then toward the end of October we knew we would have the moon to take its place. A three-quarter moon was circling above the horizon around the clock.

Before the sun disappeared entirely Gil decided to get his watch out and set it by the compass. When the sun was due south he assumed that it was noon.

"We'll need a time reference through the dark days ahead," he rationalized.

We looked forward to the winter to be shared with wildlife. Ptarmigan in winter white foraged in the low willows, daintily plucking buds from the low shrubs, conversing in low cooing notes.

Morning and afternoon the sun still shone on us, but we had "evening" at midday, when the glowing orb passed behind Noon Peak. We went out to absorb the slanting rays as much as possible. Shadows lengthened. We observed the change

A half-mile from the cabin we found warm springs. Ouzel Springs flowed all winter.

over the landscape thoughtfully — changing plumages, changing fur coats. Meanwhile Gil and I kept adding more clothing, as we hiked out in all directions every day. Snow piled up on the peaks. Long hours of sunset colored every day.

Wolves howled in the distance. Then closer came a long drawn-out high-pitched wail, like a spirit in agony. The Old Mourner was back.

As the days grew shorter we wanted desperately to use every available daylight hour for hiking. Gil was obsessed with photography.

Small straggling flocks of birds hurried south. A rusty blackbird foraged at the edge of the frost-covered stream bank. He'd better hurry. He was late. Had he waited too long?

The sun was going, going, going. On October 31 it shone for a few minutes, then it sank below the horizon. We watched the shadow creep up the mountainside.

A half-mile from the cabin a warm spring flowed out of a bank. Its temperature remained above freezing all winter long. Frost-feathered branches of willow brush bordering the stream enchanted us.

Color went from the land and into the sky. Spruce were no longer green with brown trunks, but varying shades of gray. The heavens blazed with sunset color.

"By golly. There's a water ouzel!" The chunky gray bird foraged in the stream, hopped on the frosty bank, saw a tiny fish and dived into the water after it. In the gathering darkness, Gil tried frantically to photograph the bird.

"The sun is gone." The photographer was in anguish. "I had to find it after the sun was gone!"

"I hope the little guy sticks around. He has to have open water to find food. When this stream freezes he'll have to leave."

"Maybe he'll come back in the spring."

Then we spotted two ouzels. Gil gave up on photographing them and concentrated on observing their behavior. Tiny fish occupied the shallow stream. We saw one of the bob-tailed birds swimming underwater after a fingerling.

"We'll call this Ouzel Springs."

On the way back to the cabin a flock of white ptarmigan exploded out of the low willows. Whirring wings took them to the other side of the gravel bars.

Gray jays visited us every day, taking the scraps we offered. These interesting birds belong to the Corvidae family, the most intelligent of bird groups, which includes crows, ravens and magpies. We appreciated these winter companions. The land would not be lonely as long as we had wildlife.

Frost coated the driftwood on the riverbanks.

5·Christmas in the Arctic

As winter closed in each wildlife observation was more meaningful to us. Occasionally a lone caribou passed through. The solitary ones were timid and would flee at the first sight of us, tails held high, like white flags. My heart went out to them as I wished them well until they could again find their brethren. Herd animals are not comfortable alone. There was safety in travel with the herd.

A brooding silence deepened over the land as we walked over the frozen river bars. We were apprehensive; we wondered what arctic winter was really like. Winter seemed to be preparing for a death-grip on the land, ominous — colder — colder. Deeper, darker the shadows grew. Silence ruled, silence such as we had never known. A slight breeze stirred to remind us that we were not dreaming. The scene was stark, bleak, colorless.

Clouds over the peaks lowered until snow fell in the valley. In the half-light wolves wailed mournfully to the north.

November 2. *The Old Mourner joined in and cried his heart out. Young wolf voices yapped and wailed and there was a clamorous "Eee-you, eee-you, eee-you" like a police siren, wild and eerie. It was a wild and scary chorus to come out of the dark woods, so close to camp.*

"They're howling from the same area again. It seems like that's their favorite howling place."

The next day Gil had an idea.

"Let's try to find the wolves' tracks in the snow and figure out where they were howling yesterday."

Trying to keep oriented toward the "howling place," we cut through the woods toward the North Creek.

"Here's a wolf track. Here's another." They led toward the river. Several wolves had headed in the same direction.

"Here's where they howled. Sitz marks!" Wolves had milled around here.

"You know what?" Gil sounded incredulous. "That wolf carcass is here. Under the snow.

A brooding silence deepened over the land. Silence ruled.

They're still mourning that dead mate and mother." He kicked down into the snow until brown fur was exposed.

"This is the third of November. It's three months since that wolf was shot."

"Let's follow their tracks to see where they went from here."

The wolf tracks led us to the river bars and downstream for a quarter of a mile until they came to our trail across the ice of the stream. There the pack had stopped, milled around a bit, then turned and gone up the big river.

"Look," Gil pointed out. "They went from the dead wolf directly to our tracks. But they would not cross our tracks. They sniffed them and went

the other way. They must associate us with the killing."

"You can't blame them. We were yelling and screaming at them before the wolf was shot. I wish there was some way we could communicate with them to let them know that we wish them no harm."

"I wonder if the Old Mourner will ever take another mate."

The temperature dipped to 20° below zero. In our thick down parkas, we made special trips to check on how the water ouzel was faring. The bird had moved north a couple of miles to where a spring was keeping a narrow opening in the river ice. In the deepening darkness, it looked so small and lonely on the snowy ice at the edge of the open water.

In November, the temperature plunged to 30° below and I worried about the ouzel. In mukluks and thick down parkas, we trudged north, to where we had last seen the dipper. He was not there, but upstream another half-mile we found the ouzel.

"What a durable little bird." Gil was full of admiration. "Why doesn't the water freeze right on his feathers?" The bird hopped out of the water onto the snowbank.

"The water rolls off his feathers. He's dry as can be the moment he is out of the water — an incredible little creature."

The dipper flew a few yards, dropped into the stream and disappeared under the water at the edge of the snowbank. A moment later he came

We brushed snow off a log on the ice and had a picnic at 30° below zero.

out with a tiny fish, flew up on the snow and began to beat the fish against the ice.

Gil and I kept on traveling upstream. We brushed snow off a log on the ice for a place to sit and eat our picnic lunch, which I carried. For extra insulation, our thermos and canteens of hot coffee and milk were encased in several wool socks. All the liquid was wrapped together in wool sweaters with the rest of our lunch, pancakes and fried moose meat, to keep it from freezing. We removed our heavy mitts only briefly, to finish off with raisins and nuts. Lunch over, we headed upstream.

On this mid-November day, we were surprised to find out how much water was still open. A number of warm springs were still flowing.

"It's a relief not to have to carry the rifle," Gil remarked. His pack, loaded with much camera equipment, was heavy enough. But he was used to the weight now and carried it effortlessly. The rifle was cumbersome and interfered with his pack. Now that grizzlies were in hibernation — they seek their dens in October — he had no need for the rifle. He had not only more freedom of movement, but less weight to carry.

In our arctic clothing we were comfortably warm and enjoyed our outing.

"I see some big tracks in the snow up ahead." Gil quickened his step. "Not moose nor caribou, I know that." Then he was looking down at them. "Just what I was afraid of. Grizzly! And they're very fresh."

"I thought you said the bears are supposed to be asleep by now." We looked all around.

Just what we were afraid of— grizzly!

"Sometimes, when they're hungry, they stay up longer. Evidently this fellow is late getting to bed." We studied the tracks. "He's a big bear. We ought to call him Big Foot."

"We'd better head back. How far to the cabin?"

"We've come about five miles." The bear's tracks led in a straight line toward some steep cliffs in the distance. "I sure would like to know where his den is."

"Some other day," I urged. "When you have the rifle we can follow his tracks. Let's head for home."

"Let's follow his tracks," Gil paused, ". . . backtracking," he added with a grin.

We tracked the bear for a couple hundred yards, to determine that he had come from across the river. Then we headed home, looking back over our shoulders.

A few days later we headed north again. Gil couldn't get that grizzly off his mind. This time he carried the rifle. There had been no new snow so we found the tracks and began following them, impressed with how they seemed to head in a direct line toward some likely looking rock caves in the distant cliffs.

"When the ground is frozen hard they can't dig a den," Gil informed me. "If they haven't already prepared a den in advance, they hole up in a cave or under logs and the snow covers them."

We proceeded in an absolutely straight line, following the bear's tracks, over logs, down embankments, through brush, for about a mile. Suddenly the bear tracks veered left. On hands and knees, through a tangle of alder brush, we followed his wandering route in the soft snow. We climbed over logs. When the bear waded a flowing stream we crossed on a log. We followed his tracks along the opposite bank, back across the stream, through more thick brush, until we were not far from the original deviation from the straight line of tracks. Then the bear's tracks resumed their original direct line toward those same cliffs.

"That's weird! That bear went in such a straight line until he decided to take that crazy route back and forth over the stream and through the brush, then return to the direct line again."

"I've heard of a fox doing that to throw a pursuer off its track. But I've never heard of bears doing that." Gil was puzzled. "Did that bear deviate from his straight course intentionally, or was he distracted by something?"

That winter day was short; it was getting late and we were still a long way from the cliffs. We gave up trying to find the grizzly cave and headed home, vowing to try to search those cliffs the following spring.

Night had fallen and we could scarcely see our way down the bear trail to our cabin. With sighs of relief, we pushed open our heavy door, stooped and stepped up over our stumbling block threshold to enter our cozy cabin.

Every day we donned our warm arctic clothing and went exploring, checking on the ouzel and reading wildlife tracks. Now that the sun was gone the snow was not white. Reflecting colors from the sky, it was steel blue or gray or pink. We lived in a pink and blue world, beautiful beyond belief.

Until early December we could still get water from the spring near our doorstep by chopping deeper into the ice. Then it no longer flowed.

"From now on we'll have to get water at Ouzel Springs."

At 40° below, Gil chopped a hole in the river ice to try ice fishing. I guess he didn't know how. The line froze into a solid rod.

We used the few hours of twilight at midday to go outdoors. In the cabin we burned our kerosene lamp almost continuously, from about seven o'clock until we retired for the night at about ten o'clock. A small kerosene lantern hung at the head of our bunk for reading in bed, and it supplied the light necessary for outdoor chores. To save matches Gil made a practice of lighting a candle first and from it lit a lamp and the fire in Old Lucifer, his name for the barrel stove.

There were times when we miscalculated, fed too much wood into our heater and it became dangerously red-hot. We had some panicky moments wondering if our house would burn down. When Gil was baking the heat in the cabin would drive us outside to cool off, even though the door was wide open and the temperature below zero outside. So Gil blamed it facetiously on "that devil, Old Lucifer."

In the extreme cold sounds carried clearly long distances. In the brooding silence we often heard a loud explosive sound, like a high-powered rifle shot. It happened when thickening river ice bulged up and exploded from extreme pressure.

Every morning, while I finished housekeeping chores, Gil bundled up, took his pack with camera gear, well-wrapped in heavy wool socks or sweaters, and went out hunting with his camera. He always told me exactly where he was going. Later in the crystal clear air I could hear his footsteps as he crossed the gravel bar on the big river, a half-mile away. He sounded 50 feet away.

When Gil and I went out into the deepening silence, into the winter enchantment, a deep peace filled our souls. There was so much beauty. Its effect on us was euphoric. It would have been sacrilegious to speak loudly. We were in an immense outdoor cathedral, and it seemed proper to speak in whispers. We were under a spell.

One day, near zero degrees, we traveled north on the river bars. While crossing an ice bridge over a still-flowing stream Gil broke through. He scrambled to safety across tipping plates of ice. Ice formed immediately on his mukluks, preventing water from seeping inside. His feet remained dry.

Studying the gaping hole in the ice I wondered what to do.

"You're not as heavy as I," Gil assured me. "You'll be all right. Just go to the left. Tread lightly — and move fast!"

Soon I was safely by his side.

One night the wolves presented a particularly impressive concert of many voices, like a full chorus at an opera, which continued for a long time. They seemed to be only a quarter-mile away. Next day we went out to study their tracks, to figure out how many had been singing. Out on the bars there were no wolf tracks in the snow. We

At 40° below zero Gil chopped a hole in the river ice to try ice fishing.

circled around and searched the woods on both sides of the river. Still no wolf tracks. No wolf tracks anywhere. It was as though only ghosts could have given that serenade without leaving tracks.

Gil and I returned to the cabin, stoked the fire in Old Lucifer and bundled up for an extended hike. We were determined to find those wolf tracks. The general direction was the only clue. We set out up the river.

We trudged along for several miles, following a new channel, before we found out where the wolves had assembled. First, we found a "scent post," a small cluster of short willows, which the carnivores had marked and scratched back with their hind feet. Tracks then led to their concert stage on a frozen beaver pond. Myriad wolf tracks, right below the ice-locked beaver lodge, marked the place where they had sung their wild songs. Several wolves had climbed to the top of the beaver fortress and sniffed at the air vents; that must have struck terror into the hearts of the beaver. Or, did the large rodents sit confidently inside knowing their stronghold was impenetrable? They probably didn't appreciate that wild music as much as we did last night.

How many wild ears were listening to that choral rendition? The wolves' auditorium had been the entire valley for miles around, with perfect acoustics provided by the frigid air. Meanwhile, the beaver heard it all at close range — from beneath the orchestra pit.

By the light of the full moon, a couple of

Gil and I went out in the winter enchantment. In the deepening silence peace filled our souls.

contented explorers wended their way home, slogging through the snow, toward a snug little cabin in the wilderness.

December 23.   *-12° when Gil got up. 52° in the cabin and log still glowing in the stove from last night. Moon still shining over the N.W. horizon. Moon dominates our days, since we have no sun. Temperature had risen to -6° when we hiked the one-half mile to Ouzel Springs to get water. I carried four plastic gallon jugs while Gil carried a five-gallon and a three-gallon container.*

Our trail to the spring is a foot-deep trench in the snow. Redback voles have their burrows there. We saw small holes on either side of the snow wall, with runways across the bottom of the 'canyon,' which is our trail. Delicate droppings, like small black rice, lay scattered along their pathways, with a few yellow urine stains. We stepped carefully to avoid disturbing their runways.

"Take off your parka hood and listen to the wind," Gil smiled.

"Can that be the wind?"

"Yes, it's the wind."

It blew from the north, so we were protected there. There was a rushing sound, a moaning, then a muted roar, as the wind blew through the spruce tops and roared over the ridges. Snow was blowing in plumes off the summits of the peaks.

The wind could not make up its mind which way to blow. First, it sent soft pink clouds scudding south over the peaks. Then it switched and gray sullen clouds scurried to the north. It was a chinook wind

and when we returned to the cabin with the water it was 7° above.

Clouds still chased each other north across the peaks in the boisterous, playful wind when Gil was sawing frozen meat at the outdoor pole table. The tall, slender spruce seemed to be chuckling as they swayed, getting into the game. They waved their arms, plummeting big blobs of snow on us. Gil grinned, covered his head for a moment and kept on sawing, as loose lumps of snow plopped down on him and all through the forest.

Chickadees flitted up to where Gil was working, impatiently waited for more meat crumbs from the saw, then sought shelter from another snow bombardment by darting into thick branches.

In the darkness of the afternoon, Gil baked Christmas cookies while I played on the harmonica "I'm Dreaming of a White Christmas."

On December 24, cloudy weather made it so dark that we could hardly see the two chickadees feeding on the pole table. Because Gil had better night vision, I followed him closely when we went to the spring to get water.

We prepared for our Christmas Eve celebration in the wilderness. Our tree was a shoot from a spruce which I decorated with bright cardboard cutouts made from a margarine package and finished off with some gold and red rickrack. Gil prepared the food while I put a clean pink towel on our table and decorated the cloth with spruce sprigs and cones. Then I lit a candle and we sat down to feast on roast canned ham, candied

Our Christmas tree was decorated with cardboard cutouts and gold and red rickrack.

powdered yams, hot buttered dried peas and orange Jell-O salad. Mince pie was a dessert.

After dinner I read aloud the story of the first Christmas from the New Testament. On my harmonica I played holiday music. Then we sang "Joy To the World," "Silent Night" and every other Christmas song we could think of.

We even had presents. Gil gave me a rolling pin, hand-carved from a paper birch log. I made him a special, small lampshade to shield his eyes when reading. I had also baked a special treat for my husband, in addition to the mince pie. I went to a lot of extra work to make a raisin-pecan sour cream pie. It was beautiful when I decorated the top with pecans. We planned to save it for later.

To top off the evening's festivities a little ermine came to our window and tried to get in. It was so small we presumed it was a female. She tried to catch my finger as I traced it along the window pane. We named her Yule.

Christmas Day. *No pink in the sky today, but very high clouds served as a giant reflector to give us a couple hours' daylight. We celebrated by taking a walk in an empty landscape to visit the frozen-in beaver downstream. We heard a raven in the distance somewhere. Soon we were standing in front of the beaver lodge. No sign of life. Under the thick ice and insulation of snow were at least two beavers. Their frozen mud and stick home was a strong refuge. A wolf track from upstream led past the lodge within a few feet. The canine had not hesitated as he trotted on past.*

We took close-up photos of each other for a Christmas remembrance. It was about 20° below zero.

It was about 20° below zero, cold enough to frost Gil's whiskers, when we took close-up photos of each other for a Christmas remembrance.

"It's getting dark. We'd better head back home." Gil led the way, breaking trail. We cut across to the big river slough, found several hundred yards of open water and saw wakes from several fish swimming in shallow water.

"That must be an ouzel." Gil pointed to what appeared to be a gray rock. Sure enough, it was a dipper. We slid down to see it and the doughty bird flew up and away, downstream, uttering a slight, grating call.

Ptarmigan trails zigzagged in the snow through the short willows. Ermine, wolverine and more lone wolf tracks followed the stream. At our feet we noticed minute animal tracks, like fancy sewing machine stichery, weaving through the low shrubs.

"Could be arctic shrew," Gil mused, "It's one of the smallest living mammals — weighs about a quarter-ounce."

As we proceeded home in the increasing darkness, we heard a small plane. Then we saw it coming from the North Slope.

"That's Daryl's plane. But I doubt if he can see us."

But he must have distinguished our dark figures as he zoomed over, slightly to one side of us. He waggled his wings. Although he couldn't hear, we shouted, "Merry Christmas!" and waved.

A couple of days after Christmas, we decided to indulge in the raisin-pecan sour cream pie. Our mouths watered in anticipation of this special dessert. But when I served it I detected a peculiar smell. Gil was the first to take a bite. It was *his* Christmas present.

"This tastes like rotten onion. Terrible!"

I sampled the dessert. Gil was right. It was really bad. Crestfallen, I recalled that my sour cream sauce mix was for vegetables.

"It's a shame to waste all the calories in this pie with all those pecans, raisins and dates. What should we do with it?"

"Perhaps if we freeze it the taste won't be so noticeable." We set the culinary disaster out to freeze and that helped screen the obnoxious flavor. Still it was like taking bad medicine each evening as we dutifully forced down a small portion at supper. Gil called it Fiasco Pie.

•

One evening, as a matter of curiosity, Gil decided to check our pulse rates and was surprised to find that both were down to 48 heartbeats a minute. Before entering the wilderness, our normal pulse was around 70. Gil attributed the slow rate to lack of tensions in wilderness living, added to the fact that we were lean and active.

The moon rose over Ram Peak.

6·A Time of Darkness

December 28. It was about zero degrees, with heavy dark clouds overhead, one of our darkest days, with a few hours of half-light. We went out on our snowshoes and soon were swinging into our stride.

We heard the "Ssh-ssh-ssh" of a raven's wings in the darkness overhead. I followed Gil's shadowy figure, shuffling down a trail I could scarcely see. It was snowing. All the peaks were obscured, except Mount Noon, whose top was in clouds. We could see ghostly crags at the base of Mount Ike. Gray skies reflected down to the pearly gray snow, with very flat light that made it difficult to see contours in the snow. It was the kind of light that deceived us into thinking the terrain was smooth and level. Suddenly I found myself tumbling down a four-foot embankment. I got to my feet and we waddled down the trail, straining our eyes to look for any animal tracks in the dim light.

Returning from Ouzel Springs the scene was black and eerie. Wind moaned over the ridges high above. Slanting, wind-driven snow shrouded Mount Ike. So bleak a scene! Gil plodded ahead, an Eskimo silhouette in fur-trimmed parka.

Back near our cabin, going down the bear trail, we heard a loud tapping on wood. We reached for our binoculars, which hung on shortened straps inside our parkas. In the semidarkness we could see the male three-toed woodpecker, with his golden "beanie" cap, foraging low on the side of one of the Portal Trees. We had not seen him for many weeks. Industriously he poked and jabbed, prying off bits of bark, to uncover a grub. He hitched up the tree trunk a few inches and bits of bark flew this way and that. Flakes of bark on the snow at the base of the tree proved how hard he was working. Unafraid, the woodpecker let us pass within a few feet.

Later we were inside the cabin when we heard a series of bell-like tones from the woods.

"Get the flashlight," Gil called excitedly. "That may be a boreal owl." We rushed out.

Then we heard a series of nine or ten notes clearly, like muted xylophone notes, all on the

Returning from Ouzel Springs the scene was black and eerie. Wind moaned over the ridges high above.

58

same pitch, repeated in rapid succession, come out of the inky darkness.

"It's near the bear trail."

We heard three more series of calls. Big snowflakes drifted down while we moved slowly toward the sounds. Then no more calls came.

"He must have flown away." Gil shined the flashlight around in the trees. No owl was visible. The thermometer read zero degrees when we went back inside the cabin.

About five o'clock the next evening we heard the bell-like calls coming out of the velvety darkness. We ran out to look for the little bell ringer. Again, we were disappointed.

Back inside the cabin, I checked the bird books.

"It's got to be a boreal. Peterson's *Field Guide* describes the boreal's voice as like a soft, high-pitched bell. That's it!" We were eager to get him on our lifetime bird lists.

"This book says they are tame. They should be photogenic." My photographer husband beamed with anticipation.

Gil baked bread that evening and Old Lucifer was going full blast. We kept the cabin door open and stepped outside now and then to cool off. It was a relatively warm evening, nine above.

A couple of days later, a chinook wind blew out of the south and the temperature rose to a balmy 25°. The snow turned mushy, then at night formed a hard, frozen crust, which made painful traveling for the deer. We saw some deep tracks of caribou and moose, stained with blood.

December 31 was a mild day. When we went to the springs for water at midday, heavy clouds hung overhead so we could hardly see the trail.

For our New Year's Eve celebration Gil served tender sirloin moose steaks with rice. After the dishes were washed we drank hot cups of orange spice tea and played a few hands of rummy while a great horned owl hooted solemnly from the dark woods. Later we took turns reading aloud until midnight. We wished each other a "Happy New Year," blew out the little red lantern hanging at the head of our bed, and snuggled down to sleep.

It snowed on New Year's Day and the temperature remained a mild 24°. With the door and a window wide open, Gil cooked up some moose meat on Old Lucifer.

We could hear the jays outside contentedly conversing with each other and even singing a few soft notes. They seemed to voice their appreciation of the meat scraps and the mild weather. Chickadees cheeped softly and flitted around our open window, pecking at our offerings of meat and suet tacked on the sill, while keeping out of the way of the jays, who seemed jealous because the sill was reserved for chickadees.

On a walk out to the river we met a raven flying overhead. I wondered if it would respond to singing as do so many of the other animals. I stopped, threw my head back and began singing rather loudly to him. The raven wheeled and flew circles over us, uttering a musical, liquid gurgle.

"The raven is our largest songbird," Gil laughed, as he watched it fly upriver.

That snowy day we were content to cut and carry in a good supply of firewood, stacking it to the ceiling near the door. When complete darkness overtook us, we worked indoors, updating our records of wildlife observations.

It stopped snowing after supper. We stepped outside in the darkness just in time to witness a splendid aurora borealis. Whirling shafts of bright light vaulted across the night sky, like rippling satin ribbons waving in the breeze, from west to east, all parallel to each other. Spinning circles of bright light, pale green and yellow, shot out like giant yo-yos being spun out from the northwest, leaving wide streamers behind all the way, sometimes moving forward slowly, then faster, faster, lighting up the heavens and the ghostly arctic mountains below. We gazed up in wonder at the celestial display. It was as if gigantic fire hoses shot out colored streams of water. Powerful search lights beamed across the heavens, penetrating from horizon to horizon. Twisting banners of light spread like wildfire, at times speeding across the sky as fast as lightning, now and then stopping to whirl, undulate and twist, like shimmering silken curtains, delicate and sheer, occasionally like dancing flames leaping, tortuously bending.

For a half-hour the display continued. Then it faded away. The show was over. But it left us in a trance as we entered our little cabin to retire for the night.

January 2. *Temperature dropping as we set out on snowshoes up the big river to make a new trail.*

Wind has swept away all animal tracks, to leave a hard rippled surface. An immaculate scene. Wind-driven snow, virgin mountains, with the sky a rosy purple, like wine, over the "V" of the pass to the north. No other color in the solemn sky. No wind. No sound, except the soft squeak of our snowshoe bindings. Spruce sentinels on either bank seemed to be watching us as we tramp on the flat snow-covered ice of the river. North edge of arctic tree line is a precise horizontal line on the south slope of the mountain to our right, with formidable, frozen, vertical cliffs towering above. A nearly straight line of tracks, going downstream, shows where a lone animal has gone his solitary way, sinking into the fresh snow six to eight inches with each step. We look down into a track to see the pad marks of a wolf.

Gil carried a pole to test the snow-covered ice. He stopped when it had a hollow sound and prodded a few times. It didn't sound good. He moved to the side and tested again. Crash! Where he stood, a large section of snow-covered ice gave way, slanting down. All at once, Gil was on great breaking slabs of ice. As he turned back another section broke off. Hampered by the snowshoes, he was thrown off balance as he tried to turn and back off in the deep snow on the ice. He turned clumsily on his snowshoes as more ice broke under him, then scrambled off to stable ice on his hands and knees. A yawning hole remained where Gil had been standing. We decided to go farther downstream, where we both crossed swiftly and safely, then headed back upriver.

"Look at the water smoke." Gil was using an Eskimo term for the steam, rising from open water ahead. The light was fading, but we wished to check and see if the ouzel was there. The wolf tracks led the same way. Near the steaming open water, we found the dipper, in about the same spot where we had seen it a few weeks earlier.

We stood close by watching the remarkable two-ounce water nymph chasing tiny fish under water. He emerged and round beads of water rolled off the sleek, slate-gray feathers of his back. He held a small sculpin (a fish with a large head adorned with spines) firmly in his beak, lay it down at our feet, picked it up and beat it on the ice. The bird picked up the miniature monster, adjusted it in his beak and gulped it down headfirst. He sat on his feet in the snow to warm them and watched us awhile.

"Look at his white eyelids." I knelt to get a close view.

"Yeah." Gil was observing him, too. "Those are his third eyelids, called nictitating membranes. They are thin and transparent and operate from side to side to clean the eyeball like a windshield wiper. They also act as contact lenses to protect the eyeball under water."

"Wonderful. But where did you get all that information?"

"I've been reading Welty's *Life of Birds.* Furthermore, did you notice that usually birds raise their lower lids when closing both eyes? Their upper lids are used only when they wish to wink, or close one eye." Gil and I both found ornithology a fascinating science.

That evening Gil and I witnessed another aurora, a strange one, a single arc of white light to the northeast, with a very dark curved stationary shadow of earth below. The sky was lit brightly just above the dark shadow and faded out gradually above.

"I've heard of 'ice blink.' Do you suppose this is it?" Gil wondered. (Later, Frank Heyl verified for us that what we had observed was ice blink and we were fortunate to see it. In 20 winters in the Arctic, Frank had witnessed the phenomenon only once. It is mentioned by several early arctic explorers, including Knud Rasmussen and Peter Freuchen. It has been described as a yellowish white glare — a

Water smoke (the Eskimo term for steam) rose from the open water.

white glare — an eerie glow — on the underside of clouds, resulting from light reflected by a large surface of ice and snow.)

It was a week later, at five o'clock, totally dark, when we heard the muted xylophone notes again. Slipping into down jackets and reaching for a flashlight, we stepped quietly down the bear trail toward the bell-like tones. The calling stopped. We stood on the trail waiting, listening. Gil squeaked like a mouse.

"Did you see it?" Gil whispered. "It has a two-foot wingspread." He squeaked again. Once more dark wings floated over us. Gil sent the flashlight beam searching through the nearby trees. There, perched 15 feet up in Snickaree's tree, right below the red squirrel's nest, was a little gray owl with a stubby tail.

"Oh," I murmured, "he looks so soft."

"It's a boreal, the smallest owl of the Arctic."

The little fellow was not bothered by our light or by our moving around beneath his perch. He swiveled his head around to a remarkable degree when he was not looking down at us.

We admired him until the cold penetrated our clothes. When we went back to the cabin we heard him call again.

"Let's call him Boris," Gil suggested.

"It's wonderful how a little bird can give us so much pleasure, just by letting us see him."

A great horned owl hooted in the distance.

The rest of the evening was taken up when Gil decided to develop a roll of black-and-white film. It was not difficult to darken the room.

On a cold clear morning in early January, we took an exploratory trip up South Creek on the ice. The mouths of North Creek and South Creek are directly opposite, providing a crossroads for wildlife traveling the riverbeds. A spot of sunshine on a distant high peak at the head of North Creek brightened our morning. We turned our backs on it to follow moose tracks to the big river.

Snowshoe hare tracks crossed the riverbed with tiny shrew, vole and ermine. All are vulnerable to the great horned owl when crossing a wide expanse of snow. A ptarmigan had alighted in shallow snow over ice, so it walked a few yards and burrowed into deeper snow to roost for the night. Neat brown pellets lay in an empty tunnel. Wing marks in the snow told the story of where the bird had alighted and taken off next morning on the opposite side.

Overflow ice covering the narrow canyon floor of South Creek made slippery traveling. Mount Ike towered over us to the north, turrets and parapets of layered sedimentary rock accentuating its skyline. We turned around to look back at Ram Peak and for the first time saw it in its entirety. Frequented by Dall sheep, Ram Peak is joined by a long ridge to another peak to the south, which we called Ewe. Between the two peaks is a lower prominent butte, jutting out toward the valley. The lower, we dubbed Lamb Butte. It is connected to the two parent peaks by the Umbilical Ridge.

"Let's climb up there next spring," Gil offered. "We'll walk on the tundra again and see what birds are nesting up there."

"I'd like that." Tundra has a special fascination for us. We have spent four summers on the tundra at the crest of the Brooks Range, studying its flora and fauna.

In mid-January we became aware that our food supply was going down faster than we expected. For our year in the wilderness we had brought in over four times as many provisions as we had used during an arctic summer. With the abundant supply of moose meat added to that we calculated that we would have more than enough to sustain us for a full year. So we did not skimp on food.

January 15 we took a quick survey of remaining supplies and were shocked to find that we had devoured so far more than half. It was time to tighten our belts. By eating less and not wasting food we would have plenty to eat until we left our wilderness home. Our plan was to start floating back to civilization on the Fourth of July.

During those dark winter days we went out every day, snowshoeing over the snow-covered ice of the rivers. Our trails went out in four directions. Sometimes we had only a couple of hours of dim light and at times I could not see the trail at all, but I followed close behind Gil's shadowy form. He has the eyes of a cat.

Our trails packed down smooth, becoming firmer each time we went over them. They were our highways and the wildlife learned to use them. A few inches of snow each night usually did not hide the tracks. We became adept at feeling for our pathway beneath as much as seven or eight inches of snow. When our snowshoe webs sank too deep in soft fluff on either side we knew we were at the trail's edge and would right our course.

Not once in those darkest of days did we get cabin fever. Not once did we wish we were somewhere else. The only problems were simple ones of survival and easy to solve.

Our cabin exceeded our expectations for comfort. It was easy to heat with the barrel stove. We had all the necessities of life and much more. We found beauty and peace such as we had never known before.

Sometimes when we walked through the snowy woods we would be startled by a spruce grouse exploding out of the pristine drifts, then another, and another. When it was 30° or more below these birds plunged in, tunneled under for a way, and thus they were snug when snow came down to add to their blanket of insulation. It was unnerving to have them suddenly erupt near our feet. Ptarmigan also roosted under snow cover, usually not under spruce but under the shorter varieties of willows, leaving little clusters of light brown pellets where each bird had spent the night. Grouse pellets were green from the spruce needles they had feasted on.

We never tired of our snowshoe treks. There was always something new and wonderful to experience.

We watched overflow ice forming. Springs which flowed all winter could not flow under ice frozen solid to the riverbed, so they flowed on top, spread out, and quickly turned to ice, building up layer upon layer.

Usually we traveled without speaking, listening for wildlife, watching all around. But one day we were returning home in fading light. It was 30° below. Only a couple of inches of new snow was on our trail and a firm base to travel on made easy going. We found a comfortable pace and, keeping time with our strides, began to sing.

"Over hill, over dale, we will hit the snowy trail. And our snowshoes go swishing along!" What pure joy! We grew too warm, so we unzipped our down jackets.

"No sneaking up on wildlife this way, but it's fun!"

During evenings in the cabin, Gil and I had a time reserved for "visiting" to discuss our observations and the experiences of each day. This time was Gil's idea and worked very well. We believe we saw and heard more on the trail by quiet traveling and the evening's conversations were never dull.

Traipsing along on snowshoes at 40° below, with no wind, we were too warm in our thick down parkas. We threw back our hoods and removed our mitts, going barehanded. But at 20° or 30° below, with a wind, the cold was piercing. Then we had to use knitted wool balaclavas under our down hoods to keep warm. At times, we even held our mitts over our faces to keep our cheeks from freezing.

At 40° below Gil kept his mitts on while filling the water jugs at Ouzel Springs. Water in the jugs did not slosh long, but crystallized by the time we carried it the half-mile to our cabin.

With frosty whiskers and eyebrows, Gil worked at sawing frozen moose steaks. We noticed that when the jays and chickadees came for the meat crumbs left by the saw they also sported frosty "eyebrows," a curved line of frost over each eye. The gray jay has a special feather structure which helps him endure extremely low temperatures. But we noticed that he too, like the water ouzel and the chickadees, sat on his feet now and then to keep them from freezing.

When it was 50° below outside, frost buttons formed on the nails holding our door together. The metal conducted the intense cold through to the inside. Five-gallon tins, used for food storage, froze fast to our log walls. Frost formed on the windows,

At 40° below zero Gil kept on his mitts while filling the water jugs at Ouzel Springs.

thickened and sagged down onto the sills. Gil called these formations window glaciers. At these extreme temperatures, when we opened the door a cold, thick cloud of fog, one foot deep, swept into the room, covering the floor. Still we were quite comfortable in our shirtsleeves inside. From our waists down we wore extra-heavy, wool long johns.

During the last two weeks in January and into February the cold held us in its viselike grip. The high temperature for the day would be around 20° below or even 40° below. Gil had camera problems. He wrapped the cameras in wool clothing, before placing them in his pack. Out on the trail, once he had the camera exposed to the cold, he had a chance for one or two pictures. He could not advance the film. Sometimes it broke, brittle from the cold. Then he would put the camera away to be warmed up later in the cabin.

•

Temperatures moderated on January 31 so we set out early for a long-delayed trek up the big river. We trundled upstream over hard frozen trail on snowshoes. At the first open water we were delighted to see the ouzel, like a busy little old lady clad in gray, industriously working over the rocks in the shallow stream. She rattled her little chuckle, then flew up and away downstream.

In an hour we reached the beaver lodge. Wolves had been there again and investigated the site

When it was 50° below zero outside frost buttons formed on the nails holding our door together.

thoroughly. The canines had used our trail, and had had the audacity to establish a scent post on a bit of driftwood poking out of the snow alongside the tracks. They marked our trail as their own with yellow urine stains, then proceeded to use it. Their pad prints showed how easy it was for them to walk in our hard-won pathway.

At three o'clock we started home, knowing that we would arrive after dark. Although we had over seven miles of arduous traveling ahead we were not worried. There would be a full moon to light us the last part of the way. Gil and I strode along, entranced by the lonely beauty. The crunching of our snowshoes muffled other sounds. I wished that we had time to stop and listen as we tramped into the approaching nightfall.

Then the full moon appeared over the mountain to our left. It was unspeakably beautiful. Never had we seen a moon so large. She was the lovely queen of the arctic night, sending down amethysts and diamonds to sparkle at our feet, a benevolent goddess, lighting our way home.

During the last two weeks in January and into February the cold held us in its viselike grip.

7·Lucky Friday, the 13th

 February came and the days lengthened. For two or three weeks we watched sunrise effects to the south as the sun arced below the horizon. How we longed to see the sun! Every day we gazed in fascination as light intensified on the southern horizon.

One day Gil called to me from outside. "It's almost here!" he called happily. "I can almost see the sun."

I went out with him to watch the sundog, a spot of light which followed the sun near the horizon. Wispy clouds, like shreds of white smoke, swirled over the ridge, silhouetted against the bright light, as if the mountains were on fire.

Next day, on Friday, February 13, we stood spellbound as the tiny edge of a fiery disk came into view, widened and then the sun was there in all its blazing glory. We stared, enthralled by its unaccustomed brilliance, so powerfully bright it hurt our eyes. Yet we couldn't look away.

Gil was laughing as I danced for joy. I felt like turning cartwheels. Still in shirtsleeves, I dashed into the cabin for my jacket. (It was 15° below zero.) I hurried out again.

The sun brought color back into our world. The brown cabin glowed with warm color, a brilliance over everything. A zillion tiny rainbow-colored sparkles flashed over the frosty bushes, highlighting the snow everywhere. Unreal! Then the sun was gone. The show had lasted 15 minutes.

"I can hardly wait until the sun shines again tomorrow."

It was 20° below next morning when I took a walk on the Swede Trail before breakfast. There was the loud staccato tapping of a woodpecker. Would it be a signal? It was! I heard it five or six times, about 100 yards away. Each signal was like a series of eight or nine loud taps on a sounding board. What a joyous sound ringing through the woods. Was Three-toes calling a mate?

When I returned to the cabin, I told Gil there was a new sound in the woods.

"I heard it, too," Gil smiled. "The brief sunshine of yesterday has stimulated him."

The sun was there
in all its blazing glory.
Viv danced for joy.

Every day we set out on our snowshoes. Longer daylight hours meant longer treks.

February 14. *At mid-morning we were outside for the big moment when the sun peeked over the ridge to the south. Again we gloried in the vibrant color —color everywhere! We didn't realize how much we had missed it. Little rainbows sparkled in blue, aquamarine, amethyst, bright orange, red, yellow, dancing merrily, twinkling on the snowy bushes.*

Gil and I went to walk in the revitalizing sunshine; Gil to photograph, I, to revel in it. Even at the sun's low angle, we felt a slight warmth from it. Facing the sunlight myriads of colored gems scintillated on the snow. With our backs to the sun we saw the scintillations were pure diamonds, all white.

For a half an hour shafts of sunlight pierced the forest. Sunbeams danced on the forest floor. Then no more. Everything seemed dull again.

Daylight hours lengthened by leaps and bounds. The return of the sun put a spring into our steps.

Steadily increasing sunlight stimulated the birds. Three-toes continued drumming for a mate. Gray jays, perched near the cabin, sat beak to beak, Eskimo-kissing, stroking each other's beaks, affectionately sweet-talking. Gil watched them chase another jay from the territory, then sit close together again and kiss.

We heard an answering hoot to the great horned owl's call. Two birds hooted at the same time, one on a lower pitch, out of the darkness of the woods to the east. But the most predominant bird voice in the neighborhood was that of Boris, the small owl, calling every night.

"He's calling for a mate," Gil reminded me. "Owls usually find their mates in winter." We hoped he would find a partner soon. He was really trying. His calls became more insistent, more urgent, with a half-note rising inflection. No matter what the temperature he continued. Below zero weather did not cool his ardor. On February 26 he called 150 times without a letup. We counted.

February 27. *We took one of our hand-carved benches out to south side of the cabin and sat in the sun, reveling in the golden warmth. Snow all around, but the cabin logs felt warm on our backs and we were warm. Chickadees and jays were eating scraps, Snickaree scolded in the woods. The jay called the flicker call, "Kak, kak, kak, kak." Ah, could this be spring?*

Feburary 28. *-13°. An icy fog shrouded the valley and peaks. A beautiful mysterious fog. Glancing up toward Ram Peak, we saw a rift that revealed golden sunlight on the faraway cliffs and a sparkling sapphire sky.*

When the fog lifted glittering ice crystals were dropping out of a clear blue sky, like millions of falling stars. The very air was sparkling. Crystal jewels scintillating on the snow. A gauzy fog still hung like a veil against Noon Peak. Breathtaking!

On our snowshoe trails, under intensified light, with shadows, we saw things we hadn't noticed before. We discovered a band of caribou was wintering high on the south-facing slopes of Ike peak. Through our binoculars we watched them move about a mile and a half away. Large areas were trampled and pawed, cratering where they dug down through the snow to feed on reindeer moss, a lichen, and other plants.

February 26. *14°. Gil saw caribou romping and playing on southwest shoulder of Ike. Skylined on a ridge, one ran to another, romping and cavorting, kicking up its heels. They must be enjoying the warmer weather.*

"They must be in good condition to be feeling so good," Gil reasoned. "They are still in the same area. There are tracks all over the shoulder where they have been digging in the snow to feed. I counted nine today."

What fun it was to read the animal tracks. This was our way of keeping tabs on the movements of those creatures who moved through the area without our sighting them.

A wolverine came around regularly. His distinctive set of round tracks, usually in a series of three, told us he passed first on one side of our cabin, then the other. We caught glimpses of him off and on. Gil left meat scraps out near his regular line of travel. But the little scavenger ignored our offerings. Minding his own business, he hurried past.

An occasional wolf or red fox traveled on our snowshoe trails. Lacy shrew tracks showed that the diminutive creatures were out all winter long, their ever-present hunger never satisfied. The tiny featherweights had no need for our snowshoe trails. They skittered on the snow surface, then a dime-size hole revealed where they had tunneled down. Ermine also skipped on top of the snow and burrowed under. Their dainty footprints were in pairs and the slightest crust was sufficient to support them. Female ermine weigh two or three ounces; males are slightly larger.

Our regular nocturnal visitor was a lynx, who lurked around our cabin, leaving such large footprints that we named him Big Tom. Although we caught a few glimpses of these beautiful cats in the fall, this tom was heard more than seen. His vocalizations included yowls, wails and moans, scary enough to send shivers up our spines. He stalked the hares and occasionally we found evidence where he had been successful.

Snowshoe hare trails led through the willows and under spruce around our dooryard. One day Gil ducked into the cabin with an announcement.

"I've found the most unusual tracks. Come out and see what you can make of these."

A wolverine came around regularly. He left distinctive tracks, usually in sets of three.

Snowshoe hare tracks led through the snow.

"I've found the weirdest tracks." Snowshoe hares had danced in the snow.

I reached for my jacket. Out on the snowy gravel bar, south of our cabin, he showed me a mystifying herringbone arrangement of snowshoe hare tracks.

"Look." Gil was setting up his camera. "If I hadn't seen typical hare tracks leading up to them I wouldn't have recognized the source. What do you think those rabbits were doing?"

"Hmm . . . Olaus Murie in his *Field Guide to Animal Tracks* has a drawing of arctic hares hopping on their hind toes and he mentions the strange tracks they leave. I always wanted to see that. Are you sure these aren't arctic hare tracks?"

Gil shook his head. "Too small."

"Well, these snowshoe hares certainly were dancing on their hind toes. Wonder why?"

"We've seen a number of wild species playing. Sally Carrighar, author of *Ice Bound Summer*, says snowshoe hares like to play in the snow."

"There was a particularly bright aurora last night. Do you think that stimulated them?"

"If we'd known the rabbits were having a fling we could have joined in the frolic. Sure was a pretty night!"

During the coldest weather we rarely found a moose track along the rivers. They stayed high on the slopes. One day, through binoculars, we watched a large moose climbing Lamb Butte. The moose, without antlers, strode easily above tree line.

"What are the moose doing so high on the mountains?" Gil wondered. Later we noticed that two of the beasts were spending much of their time

Gray jays with frosty eyebrows kept coming for scraps.

on the slopes of Noon Peak, south of our cabin. Their dark forms were easy to spot from our dooryard as they browsed on shrubs protruding from the snow, high above the valley. They seemed to stay in the same general area.

Gray jays with frosty "eyebrows" kept coming for scraps. One day a jay came and talked softly to Gil while he was sawing meat. Gil spoke to the bird and they carried on a long conversation.

I was reminded of Conrad Lorenz and his research on the jackdaws, another Corvidae, which he describes in *King Solomon's Ring*. One of the jackdaws became enamored with Lorenz and tried to feed him a worm. When he refused the gift the bird tried to stuff it into the scientist's ear. I wondered if Gil was getting too chummy with the jays.

The brown-capped boreal chickadees were our favorites. A number of these trusting little elves, less than one-half-ounce bits of vivacious fluff, came to our window sill with tiny frost cyrstals clinging to their "eyebrows," on top of their heads, and under their chins. With amazingly loud taps they attacked the frozen moose fat on the window. Their drawling call, "Chicka-dearr-dearrr," reminded us when they wanted a handout. They were our "Chickadears." Lengthening days had stimulated their spring song.

"Spring is here! Spring is here!" the little chickadees whistled merrily. Their cheerful message brightened our days, even if it was 30° below.

One such cold, windless day, we shuffled on our snowshoes a couple of miles up North Creek to call on one of our neighbors. A porcupine had a den in a narrow-mouthed, rocky cave, near a bend in the river. A thick patch of willows and alder separated the animal's retreat from the open gravel bars. I stooped to unstrap my snowshoes.

"Why don't you wait out here while I go check and see if Porky is at home," I suggested to my husband. He sat down to scan the surroundings with binoculars while I made my way through the tangle of shrubs to the porcupine's den.

I squatted down to take a look at bean-sized fecal pellets of compacted, greenish material, which lay below the entrance where the rodent had swept them out from his chamber, which was at a higher level inside.

I began humming a soothing lullaby tune, hoping the porcupine would become curious enough to peek out. I hummed softly for two or three minutes. There was a rustling sound in the den. Encouraged, I kept humming. More rustling sounds. The little fellow was moving near the entrance. More humming, then I listened for more rustling inside.

"Hmmmm . . . hmmmm . . . hmmm," came from within the little cave. I could hardly believe my ears! The porcupine was singing, too — all on the same note, about middle C.

I hummed my little tune, going from middle C up two notes and back down again, about five or six times. I listened.

"Hmmm . . . hmmmm . . . hmmmm," replied the porcupine. "Hmmm . . . hmmm . . . hmmm."

We continued our duet, taking turns. Each time I paused the porky sang back to me. I kept on, afraid to break the spell.

I wanted Gill to share this precious moment with me. If I called to him it would frighten the little beast. If I stopped singing to get Gil the porky might not be in the same mood when we got back. I kept on singing, hoping Gil would wonder about me and come looking. Ten more minutes I sang to the porcupine and he replied with three notes, "Hmmm . . . hmmmm . . . hmmm."

If Gil doesn't hear this he won't believe it, I

The chickadees sang "Spring is here," but night temperatures dipped to 30° below zero.

73

thought. I couldn't stand it any longer. He had to hear this! I turned and hurriedly scrambled back through the alder brush to where Gil was waiting.

"It's about time," he complained. "I'm tired of waiting. What do you look so happy about? Did you see it?"

"I've had an unusual experience. The porcupine sang to me."

Gil gave me a funny look.

"I know it sounds crazy. But he did!" I insisted. I explained the details. "Come listen to him hum."

"Well, okay." He unbuckled his snowshoes. "You do seem to cast a spell over them. But a porcupine singing to you is a little too much."

We threaded our way back toward the porcupine's den. Gil was breaking branches to make his way through the tangle of alder brush. He always required more space to squeeze through than I.

"Be careful. You move like a moose. You're making too much noise." I worried that we would alarm the creature.

"Now be very quiet and listen." Squatting near Porky's doorway, I began humming the same three notes, up the scale and down again, over and over again, for two or three minutes. Not a sound came from within the den. I hummed again. I stopped to listen, hummed, listened. No response came from the porcupine.

The animal refused to sing again. And Gil never had a chance to hear a porcupine sing.

"That darn porcupine!" I was so disgusted. Why couldn't he sing just once for Gil?

It was 30° below when we stopped to visit the

Gil found the porcupine near the top of a spruce tree, feeding on the needles.

porcupine another time. We checked his den — no luck. Finally Gil found him near the top of a spruce tree nearby, feeding on the conifer's needles. We climbed the slope beside the tree to about the animal's level. He held very still, eyeing us suspiciously. He stopped feeding. I began humming. He turned his head to look my way, then relaxed, no longer afraid. As I hummed, he reached with his front paw to pull a spruce twig closer and began nibbling.

March 9. *Minus 7° and very strong winds, so we spent much time indoors. A good time to take a precise inventory of our food supply. We find we have just enough to get by for breakfasts and dinners, but short on foods for lunches. We will begin to ration more stringently.*

Although Gil hasn't baked bread since January we are short on flour, so will have to be careful how we use it. We have figured out how many cups of each item we can allow for each week, so we won't run out. We are short of salt, but not sugar, for which we are thankful. The minerals in our water are a salt substitute. We notice no leg cramps, even though using less and less salt.

We are saving the canned meats and fish to use when our moose meat runs out later in the spring.

To boost our morale, we can manage to have one very small pie each week, using one-half cup of flour for the crust.

"Friday will be pie day," I laughed. "We'll bake then, and make each pie last four days. Fridays, we won't eat any, but we can sniff it and enjoy the aroma and anticipation. For the three succeeding days, we can each have one-sixth of the seven-inch pie." When we finished off the pie, it helped to know that we could bake another in three days.

"What we really need is more flour power," Gil commented, taking a cue from the folk in San Francisco.

•

In February, Gil and I made several snowshoe trips part way up the lower slopes of Noon Peak. Early in March we decided to climb all the way to the summit. In preparation, we had been on an extra meager diet for a few days to allow more food for the strenuous efforts required for the climb.

It was 47° below when we left the valley floor. Wind and snow had obliterated our former snowshoe trails. Out of the wind, on our webs, laboring through deep and fluffy snow, we grew warm from our efforts. A few scattered spruce near the river thinned out and we made our tracks on top of tangles of dwarf birch and some scrubby alder; only a few dwarf willows protruded. Finally we were above where any vegetation showed above the snow. The temperature moderated as we climbed. It was around zero degrees and windy on higher ridges.

"It's warmer up here than in the valley," Gil noted. "Cold air always sinks when there is no wind. That's probably why the moose were up here." The big beasts had moved to another slope before we began our climb.

Gil and I alternated breaking trail. In places it was so steep that we had to remove our snowshoes. Reaching a fairly level "bowl," we sat on our snowshoes, with our backs to the wind, to each a lunch of hot milk, pancakes, moose meat, cookies and hot tea. I also carried raisins, and peanut butter and honey in a squeeze tube. And a few candies. Because we had been using so little salt lately I took a salt tablet to prevent leg cramps.

Our goal was in view ahead, with an easy grade below, but the trail had to be made in soft deep snow. Gil started ahead to break trail. All the way up we followed traces of tracks made by a single red fox, climbing the mountain ahead of us. Usually the fox did not sink in far and had pretty easy going.

March 12. *The wind gained in intensity. Snow plumed off the ridge above. Suddenly both my thighs cramped up and I was in extreme pain with each step. Gil was above, oblivious to my distress. I knew the salt tablet I had taken would eventually eliminate the cramps. Luckily I had taken it. I gritted my teeth and kept going. Wind was blowing powder snow across where I waded into the foot-deep trench made by Gil's snowshoes. I stopped and looked up at Gil's figure, working skyward in the gale. I moved on and up. Snow was already filling the trail. I hoped the cramps would go away soon.*

Only Homo sapiens *would be crazy enough to climb this mountain in the winter, just to get to the top, I thought. Then I did a "double-take." Was the red fox going for the top, too? Why?*

In early March we tried to climb the summit of Noon Peak.

Viv took off her snowshoes when the climb became too steep.

Soon the leg cramps eased and I moved without pain. I caught up with Gil.

"Want to break trail for a while?" he grinned.

"Sure." At first I sank in over my knees, then I reached a wind crust formed on the snow and stayed on top, but I was struggling to keep my balance in the fierce wind. Gil followed closely, but broke through the crust while I stayed on top. Near the ridge, I sank in too. We persisted through breakable crust up a steep slope and anticipated reaching the top. A finch chirped and flew across ahead of us in the gale.

What a brave little bird, I thought.

"There's the summit, over to the right," Gil pointed. It was only a few hundred yards away and not much higher. A good snow crust led to it. Happily we plodded over in that direction. The summit was before us at long last!

"Let's go up together." We encountered deep snow again and floundered up the last few yards to the top of the snowy dome.

"Oh, no! This isn't the summit!"

In dismay, we saw another dome of snow about half a mile away that was a few hundred feet higher. And separating us from the true summit was a 300-foot-deep pass.

"We've come six and a half miles so far." Gil checked his pedometer.

We gulped down some candy for quick energy. I led onward, traversing the steep descent, while Gil stayed back for photos.

Fox tracks led down and then upward toward the summit ahead of me. Climbing up a ridge leading to the summit I waded through knee-deep snow. I climbed eagerly now, feeling exhilarated because we were going to make it. On the ridge we were climbing I waited for Gil and we climbed to the top together.

On the summit we were in a different world of mountaintops all around, pointed, domed, sawtooth and ridged summits. We looked down to the band of caribou on Mount Ike and saw where they had pastured all winter and dug their craters in the snow over a vast area. We could see moose on the slope below.

We turned from the wind in our warm parkas and faced the sun in a cloudless sky. Gazing at the pristine peaks all around we felt a euphoria. We were grateful that there still was a spot on earth where man had not yet proliferated. What a rare experience!

At our feet we noticed signs that the fox had been here before us. His tracks led down the other side of the summit.

Thoughtfully we wended our way down from the untrammeled heights. Why had the fox climbed the mountain? We could think of no reason which related to his survival. For fun? Far below, we could see where he had curled up in the snow on top of a ridge. Why? To watch us?

I set the pace most of the way down. We could easily follow our own tracks, but in places the wind had filled the trail with snow and we had to open it again. Downslope a ways, we stopped out of the wind to watch the sunset color blaze across the sky and I wrote in my diary.

March 12. *These precious days of being alone in the wilderness are so indescribably beautiful; we cherish every moment. The Master Painter had brushed pure gold streaks across the setting sun, behind immaculate white peaks, with touches of mauve and pink. Splendid. Down valley, Takahula peaks streaked with pink. Alpenglow on the virgin snows.*

When we reached the valley floor there was the fox, a black phase of red fox, out on the snowy bar not far from our cabin, sniffing at our tracks. She watched us for a while. We named her Melanie.

"It sure would be nice if she would hang around here for a while." Gil planned to put out some offerings of moose meat for the fox to gain her confidence.

That evening we were very tired and happy to be back at the snug cabin.

The day after our Noon Peak climb very strong winds waved the spruce on the valley floor. The meat cache creaked as the moose meat swung wildly under the netting.

"Let the wind blow. We're lucky we're not climbing that mountain today." We looked up to Noon Peak and saw snow plumes blowing off its summit. We were content to spend a lazy day resting. With our food supply low and on short rations we were steadily losing weight and tired easily.

One day of inactivity was enough. The next day was calm so Gil took me across the big river to see a squirrel's "mushroom tree," a 30-foot-high spruce, loaded with dried fungi. All over the branches, from close to the ground up to and above a round nest made of grasses, we saw brown and tan mushrooms, lying singly or in piles, tucked here and there among the green twigs. A red squirrel's fresh tracks were in the snow at the base of the tree. Previously we had seen a number of old bird nests loaded with mushrooms, where squirrels had cached them.

One morning when Gil was on a photo hike, I stayed near the cabin and heard a doglike bark from the forest to the east. Then I heard another bark.

It must be a fox, I thought. It was the kind of barking a puppy or small dog makes when it is

We finally reached the top of Noon Peak.

77

romping and playing. The barking continued for several minutes. Was it Melanie?

The fox was around, we knew that from tracks. And we had seen the tracks of two together. Probably they were romping in the snow, playing, just as we had observed at Lonely Lake at the crest of the Brooks Range in the spring of 1965, while we were doing some research for the U.S. Fish and Wildlife Service. It was fun to hear the barking. I wished we could watch the foxes play.

With more daylight hours we enjoyed more glimpses of wildlife and the increased activity as the breeding season approached. Gray jays sunned themselves and people-watched.

A cow moose trusted us enough to bed down in the snow 150 feet from the cabin. We named her Moosette. Some caribou came down off the high, south-facing slopes to move about. We found a place not far from Porky's den where they had cratered down through 18 inches of snow to feed on reindeer moss, pawing it out with their enormous cloven hooves, using a front leg. What a hard way to scratch for a living!

Not often did we stay indoors because of bad weather, but March 15 was such a day, a bleak and dreary day, with a miserable wind, snowing hard the entire day — a genuine blizzard.

Gil unraveled a worn-out sock to get yarn to darn a better pair while I worked at housekeeping chores. We watched the jays and chickadees feeding at our window sill until we became bored with the inactivity.

Then we donned our cold weather duds and we waddled on snowshoes through the thickest stand of trees in the Enchanted Forest. Even there the wind tugged at our clothes. We found a woodpecker hole about four feet from the ground, used last year, we thought from the new look of the cavity opening.

March 15. *We stayed out for only one hour and returned to our snug cabin. How warm and safe we feel here! Miserable wind all day, and snowing. Worst blizzard we've had. Snow blowing, covering all trails, except in the woods. We listened to the roaring of the wind and the creaking of the cache. We estimate about a 30 mph wind, with -4°. After that brief outing we were happy to stay indoors and read.*

A fox climbed the mountain ahead of us. His tracks show to the right.

8·Wild Romance

The most fascinating season of bird watching — of all animal watching — was fast approaching. In fact courtship time had arrived for some species.

On March 20, it was 20° above at breakfast time. We watched the feeding post. Chickadees were there. The jays were late to arrive. They pecked and pulled at the meat scraps, each jay on opposite sides. Then we saw one jay tug and loosen a piece of meat, fly a short hop to the other side and present the tidbit of food to his mate. A couple of minutes later he flew to an alder bush at the edge of the island, just across the gravel bar. He picked up some small object, which we could not identify, then flew to the feeding post to offer his companion what he had found. She took it and together they flew over the cabin into the woods.

Later in the day Gil stuck his head in the cabin. "I hear two woodpeckers drumming. Let's go out and see if Three-toes has a rival."

The rapid drumming led us along the bear trail and beyond. Raising his binoculars, Gil spotted the second percussionist.

"It's a female. No yellow on the head." Three-toes' persistent drumming had won him a mate.

The birds ignored our presence. Three-toes drummed. She replied with a more subdued series of signals, then proceeded to forage on the tree trunks, jabbing, prodding, twisting chips of bark to expose grubs and insects. She seemed to ignore the male, but when he signaled she answered, her vibrating head busy all the time.

"What amazes me," I mused, "is that they don't get headaches. It seems to me they should get punch-drunk."

We had been hearing a curious pumping, throbbing sound coming from the woods. I thought it might be a grouse.

"No," Gil decided. "It might be a large hawk."

On March 22 we struck out on showshoes for the Enchanted Forest to investigate a big stick nest which we had located about 50 feet up, near the top of a big old cottonwood. Two feet of deep snow covered the stick platform built by hawks.

Ptarmigan left their tracks in the snow. Feathered feet served as snowshoes.

"Notice anything different about the nest?" Gil focused his binoculars on the mass of sticks.

"Yes. More sticks have been added on top of the snow!"

"Yes. Fresh sticks, about a foot high. Strange that they are placed on the snow. You'd think the stupid bird would push the snow off first!"

"Let's get out of here so we don't disturb the hawks."

Two days later we noticed a pair of goshawks at the nest. More sticks had been added and arranged nicely, curved around the edge and settling into position. The snow was only about eight inches deep.

March 30. *Breakfast this morning was three-quarters cup oatmeal, boiled moose meat, one cup milk and one-half cup of coffee. We sat at the table and watched snowflakes drift down. A gray morning. Like a typical winter day in Oregon's Cascade Mountains.*

"This ought to cool the goshawks' ardor," Gil said. "It doesn't look like spring."

Today Gil and I felt lethargic. Gil is so thin I can't believe it. We don't have as much energy now that we are on shorter rations. We are eating lean moose meat three times a day. Getting slightly tired of it, but I never mention that. We are thankful to have it.

The cloudy, dreary weather affected us. We did manage to go out and saw a little firewood.

March 31 was very windy, snow depth, 23 inches. After a meager lunch Gil kept talking about food. To take our minds off of eating we went out in the wind for a snowshoe hike and immediately felt better. We noted the band of caribou still on Ike's peak. They were in a close huddle, pawing through the snow and feeding.

"Look at those ravens flying over the caribou." Gil watched through binoculars. "They're playing."

On that windy afternoon we watched two ravens playing tag and follow-the-leader over the ridge. Then, side by side, they soared in circles and in large "S" curves, wheeling around each other, higher and higher. Back they came, always keeping close together, wing tips nearly touching. We noticed another pair of ravens join the game, keeping together, as graceful as ice skaters, going back and forth, up and down, with their partners. Another raven tried to join one duo, but, like a fifth wheel on a wagon, he looked out of place and didn't fit in.

We watched until our aching arms could no longer hold up our binoculars. The ravens were still cavorting over the mountainside when we stopped watching.

Supper was dried, cooked lima beans, moose meat and Eskimo ice cream. To a bowl of snow we had added powdered milk, raw bone marrow and some of Gil's mother's strawberry jam. Delicious!

One day in early April I was in the cabin mending one of Gil's wool gloves, being very quiet. Then I heard a scratching on the back end of the cabin. I went out and discovered a jay digging into the sphagnum moss chinking. He carried some away.

Gil returned from a photo hike and received the news with a smile. "Let the jays have some. We can replace it and we'll get our reward by following them to find their nest."

Day after day the jays continued to make off with the chinking — while I replaced it — and Gil began his project of jay-watching. But it wasn't easy. When one of the birds disappeared in the woods, the photographer took his cameras and went to the edge of the woods, then sat down to wait for the bird to fly past on another trip, to see where it went from there. Sometimes, after a long wait, no bird would show up. Or, finally Gil would discover a jay was perched above, silently watching him. Very frustrating!

"I haven't talked to a single Eskimo who has found a jay nest," he informed me. "If I keep trying I'll find it."

April 5. After an early lunch of moose meat and milk we set out for a snowshoe trek up North Creek. Wind and snow had wiped out our hard-earned trail once more. Under the unmarked snow was the solid base of our pathway. We could scarcely make out a faint line here and there. By feeling the way, trial and error, we mushed on. It was remarkable how easily we could stay on it. The beautiful clean snow was unsullied except for tracks of Big Tom crossing into the Enchanted Forest. A wolverine had crossed a little farther up last night, too, leaving a three-track pattern. The dainty trail of an ermine crossed our trail. About a mile up two big moose had ploughed into our trail and messed it up.

I was tired and hungry. Our main staple food now was lean moose meat three times a day. Protein meals didn't give me much energy. My engine is fueled with carbohydrates and I was running out of steam. In a narrow rocky canyon we sat on our snowshoes to eat and regain strength. It was only 10° above zero, but the sun gave us warmth. With down parkas we were comfortable. I had carried a snack of three cookies, some candy, raisins, two figs and a few nuts, and we had a quart and a half of hot water.

While we rested three or four jays flew across the canyon and disappeared into the spruce. Two flew to another spruce. I focused my binoculars to see if they were nest building. They were perched side by side, people-watching. Our snack revived us and we were ready to move on.

"Look! Look! Look!" Gil excitedly pointed up toward the lower spruce-covered slopes of Ram Peak. From out of the blue came a swirling white cloud. A swift-moving, living cloud of white wings, flashing in the bright sunshine, swung in close to the flanks of the mountain and dropped down into the trees and bushes on a ledge above. About 125 ptarmigan in one flock, which wheeled and turned as one unit, was closely followed by another flock of about 100 more. What a marvelous sight of immaculate beauty! They came and there was a muted sound, like the roar of wind, from their strong wingbeats. Their arrival had taken less than a minute, but it left us spellbound.

We had witnessed another miracle of nature. What a privilege to witness this phenomenon of

Day after day the jays continued to steal the moss chinking from our cabin.

immeasurable beauty! There was something miraculous about the whiteness of them, their flocking and flying in one cohesive movement, acting as one living organism. They are preyed upon by so many other species, yet they endure. How grand to see them on their journey north!

We continued our journey up North Creek. Gil, who was breaking trail, stopped dead in his tracks and pointed to some low willows near us. Ptarmigan stood like white statues, heads held high, tails in the snow, perhaps to hide the black feathers in their tails. We held still and the ptarmigan eyed us for a couple minutes. Then, ever so slowly (we had to look closely to see that they were moving) the white bird statues seemed to glide imperceptibly away from us in complete silence, heads still high, tails low. Only four birds moved very carefully away from us until we could just see their heads over a low snowbank. Their heads went down and they were out of sight. Then we heard a whirring of wings as they took off upriver. A dozen more birds appeared from the steep snow slope facing the stream. These followed the others. As long as the immaculate birds held still they were practically invisible. The beauty of them!

Around the next bend a ptarmigan uttered a soft guttural "Kruk — kruk," a pheasantlike call. It flew down from a bank at our left and landed in short willows near us. We didn't disturb it. We knew it was a "frozen" statue in the snow. We went on. Ptarmigan trails zigzagged in the snow ahead of us.

After traveling about six and a half miles Gil and I turned back. We spotted two Dall rams, up to their bellies in snow, on the steep slope near some vertical cliffs on Ram Peak.

We made good time on the trail which we had made, but often stopped to look and listen. A sudden muted roaring of wings came, then about a hundred more ptarmigan overtook us and passed close over our right shoulders. They flew downriver around the next bend.

"Here come some more!" I turned. We had time to see three or four ptarmigan flying low on the opposite side of a clump of willows.

"There's a hawk after them." The goshawk swept down toward one, missed, wheeled and perched, teetering on the tip of a spruce. The hawk cocked its head and kept looking below at where the ptarmigan had plunged down to the snow and disappeared with their camouflage. We waited. The hawk was still perched there, looking down for his prey, when we left.

Traveling toward home, we noticed ground squirrel tracks where they had emerged from burrows under snow-covered rocks near the river and hopped across the snow to the opposite bank. This was the first evidence we had that the squirrels were out of hibernation.

April 6. Just after sunset Gil watched a snow-white hare hop across the gravel bar to the edge of willows on the island, where it met another hare. They began to play, chasing each other in tight circles and zigzagging back and forth, faster than the eye could follow, white bunnies on white snow, having fun. Gil called me. A moment later they were in the brush and we caught glimpses of their fluffy white fur, chasing past in the willows. Soon patches of bare ground would make them more wary and spoil their fun.

Next day a small flock of hoary redpolls flew over, calling, "Chi-chi-chi-sweet." They perched on the tip of a spruce by the cabin. We watched through binoculars as they searched for food. We saw that they were eating buds from the spruce. Soon they were on their way, flying northward in undulating flight.

A snowshoe hare lived around our dooryard. He trusted us as he hopped around leisurely in impeccable white fur coat, eating spruce tips dropped by Snickaree and alder leaves newly exposed by melting snow. Sometimes we found his dark eyes studying us for a moment, then a few hops took him to the next tidbit. We called him Snowshoe Harry.

Camera in hand, Gil patiently worked to gain the hare's confidence, moving slowly and quietly while humming softly to him. After several months the rabbit went about with no fear of us. His trails wandered all around our yard and into the woods near our cabin.

One day in early April, we trundled up North Creek for many miles on showshoes. The days were longer and so we hoped to go perhaps 10 miles up toward the crest of the Continental Divide. Passing Porky's den we went around the next bend and found the fresh tracks of an animal bounding through deep snow. It entered our trail and followed it for a few hundred yards, before

A snowshoe hare lived around our dooryard. He trusted us.

85

bounding on into the woods. The leaps left deep holes, which were spaced about eight feet apart. We decided that the lone traveler was probably a wolf, a large one. The soft deep snow made traveling very difficult for many species.

"How fast can you take off your snowshoes if we meet a grizzly?" I asked Gil jokingly.

"About two seconds!"

"I'd probably try to climb a tree with snowshoes on," I replied. The possibility of meeting a grizzly was rather exhilarating; we carried no firearms.

The dainty tracks of an ermine led over the snow. We saw the pretty little prints of their feet on the hard surface. It was fun to read the animal tracks.

A band of caribou had marched in single file from upriver, all stepping in the same tracks. Drag marks from the hooves left a narrow trench between deeper holes where they stepped. They had much easier going when they encountered our trail. Moose did not benefit from using our trails. They broke through anyway.

Gil was anxious to extend our snowshoe trail farther upriver to look for a wolf trail crossing the snowy riverbed. He led the laborious trail-breaking. He hoped such a trail might lead to a wolf den.

We entered a narrow, steep-walled canyon, which wound tortuously even higher. We were proceeding up the snowy ice of the river beyond the last trees. We faced a slight breeze so I wore my hood up. I felt the lonely silence.

"Rawrrr!" A roar shattered the stillness.

Gil and I stood rooted to the spot. I threw back my hood.

"Rawrrr!" It came again, a terrifying roar, like that of a lion in a zoo.

"What was that?" My heart was in my throat. "A mad bear?"

"I don't know." Wide-eyed Gil looked all around the cliffs on both sides of the canyon. "It's a big animal for sure. And it sounds mad."

"Rawrrr!" It was closer this time. The bellow bounced back and forth against the canyon walls. From where we stood we couldn't tell which side it came from. I moved closer to Gil.

"I'd feel better if I could see it." Gil's eyes still searched the cliff.

"Let's get out of here." Desperately I spun around, looking for a tree to climb. None was in sight. "Let's go home." Trapped in a box canyon, I felt like an early Christian in a lion pit.

"Home is seven miles away." Gil looked scared.

I waited for his decision.

One more angry blast came from the unseen beast and Gil pointed back down the trail we had just made. We headed back at a fast clip. The frightful roars continued at about two-minute intervals. Each time we managed to accelerate our pace.

"It's following us." I kept looking back over my shoulder.

"Sounds like it." Gil kept looking up at the cliffs. "I see tracks of a large animal going up — or down the slope — up there."

We stopped a moment to check through

binoculars. While we stood there another roar came! Then a coughing sound issued from the cliffs.

We took off down the trail, jogging on snowshoes. I spotted some trees in the distance and selected two for climbing . . . if we could get to them! More roars spurred us on as we dashed around several bends in the river before we heard no more. But we kept up our feverish pace for several miles.

"Let's stop for some food," I panted. Weak and trembling, I let down my pack, making sure two good climbing trees were nearby. A brief rest and some candy and milk revived us. We resumed our undignified flight.

Two miles from home we began to breathe more easily.

"I guess I'll carry firearms from now on," Gil grinned ruefully. "I wonder what that beast was."

It was good to get back into our snug, safe cabin. I climbed up into our bed and collapsed.

•

Three-toes, our bird friend, continued to drum merrily, proclaiming his territory.

Spring was surely on its way. We had seen the first migrant bird in March, a golden eagle. Now came the first migrant passerine (perching bird).

Only a few grass seed heads were exposed above the snow when the sparrow-sized snow buntings appeared. Small family groups flew overhead, emitting chirps with each undulation in their finchlike flight. A male, still in winter plumage, alighted on the south-facing bank near

our cabin to glean some seeds from last year's grass. This was the only exposed earth in the vicinity.

April 11.   *The little lone traveler was tame and let Gil get within four feet. We hadn't seen a snow bunting in winter plumage before. They change into breeding plumage a couple of weeks before nesting. It stayed a half-hour, searching, foraging on the bank. Jays were curious and soft-talked to it. Chickadees, our welcoming committee, flitted down, perched close by, lisped excitedly, inspecting the new arrival.*

The bunting hopped to the edge of the snowbank, reached up to grasp a long grass stem in its beak, then slid its beak along, pulling the stem down, until the first seed was in reach. This tiny rugged individualist dared to come to this severe habitat, ahead of his flock, gleaning what little food was protruding from the snow.

The snow bunting made this a red-letter day. 2°, bedtime.

The male snow bunting in winter plumage has brown marks on his head, back and breast; in summer his head is all white and his back is black.

Our most vocal and sassiest neighbor was Snickaree, the female red squirrel, who lived in a 12-foot stub of spruce trunk by the bear trail. She scampered into our yard and chewed on moose fat scraps. I'm sure she thought she owned the place. In the coldest winter weather she spent much of her time resting in her cozy nest. But on milder days she was out working over spruce cones,

The male snow bunting in winter plumage has brown marks on his head, back and breast.

scattering the scales on the snow below. Her kitchen middens were all through the woods. When she sighted anything suspicious or threatening, she scolded with a curious four-beat rhythm, like an old four-cylinder gas engine.

One day Gil left an odd-shaped piece of wood on our pole table in the yard. When Snickaree spotted it, she clung, head down, from a nearby tree and in a frenzy scolded the piece of wood for a quarter of an hour. Tired of her racket, I went out and removed the threatening object. Then she confidently skipped across to the moose fat tub, a five-foot spruce stub where we fastened suet for the birds.

One spring day Snickaree was playing tag with

another squirrel, racing up a spruce, performing daring aerial leaps to another spruce, down the trunk, across the ground and up another tree. "Those squirrels sure are frisky this morning. That must be her mate." Gil grabbed his camera and went out to get her picture.

The squirrels' games of tag went on for a week or more. Sometimes she streaked up a tree, and tired of their game, she stopped and turned to confront her follower, snickering crossly at him. We saw him stop, clinging to the tree trunk below her. He didn't bother her then. Just patiently waited. She took this time to grab a spruce cone and rest while eating it.

It was not long before we noticed, while she was sitting on the moose fat post, that Snickaree was getting pear-shaped. She seemed to crave the suet. Later we saw her skittering across the bear trail with a mouthful of nesting material, leaves, grass and sphagnum moss.

April 16. We wandered out to the Enchanted Forest looking for a jay nest. No luck, but it sounded like spring. The male pine grosbeak warbled a lovely spring melody. His lady answered with sweet notes. Redpolls and crossbills twittered and called while feeding in the spruce tops. Then we heard "Chickadear—rr." All the while the goshawks were calling loudly from the vicinity of their nest.

Gil and I approached quietly and stood looking up at the massive nest. No hawks were visible. All of a sudden large wings swooshed close to our heads, with loud cries of "Kyak—kyak—kyak," as a

Our most vocal neighbor was Snickaree, the female red squirrel. She thought she owned the place.

goshawk tore through the forest with frightening speed.

The message was clear. "Get the heck out of here!"

We headed for home.

April 19. *Gil and I looked up to see the caribou frisking and playing on the south face of Mount Ike. One started running slightly uphill, then three followed in single file. They were in high spirits, kicking up their heels and racing first one way, then another. What fun to see them playing! They were still on a high shelf, where they have lived since moving a short distance from the southwest shoulder five days ago.*

It was April, and Boris, the boreal owl, still had not found a mate. He was still calling. Week after week, month after month, the resolute little owl kept trying. When we hiked two miles from the cabin we could still hear his clear bell-like notes. One day I watched him as he perched about 10 feet above. He wore a grayish-white vest with vertical cinnamon stripes and a polka dot helmet on his round head. His eyes were closed, as if he were sleeping. Abruptly his eyes popped open, round as saucers. He stretched out his neck and his throat feathers puffed in and out as the bell-like tones rang out loud and clear, after which he seemed to sleep again. Moments later, he again came alive to repeat his urgent calls.

One sunny morning Gil sat on a bench outside the cabin, industriously making a birdhouse from the hollow shell of a spruce log. He sawed off a 15-inch section, fitted a flat top and bottom and was enlarging the nest hole opening with his pocketknife.

"I sure hope Boris finds a mate." He smiled. "I've got to have a picture of one of his offspring." My wildlife photographer got carried away with the project and made about eight birdhouses of assorted sizes and mounted them around the yard. The boreal's nest box was placed atop a 12-foot-high spruce stub near the cabin.

April 20. The entire day was dominated by the boreal's calls which came at approximately 40-minute intervals. By evening Boris started calling

One sunny morning Gil sat on a bench outside the cabin making birdhouses.

every quarter-hour. As darkness fell he moved to the vicinity of the nest box and called for hours on end with only short pauses.

The following day Boris continued the same routine. During daylight hours he called like an old town clock chiming out the time. At 3:00 P.M. the chickadees began scolding the owl. When the jays joined in the harassment, darting close to him, I noticed that they were slightly larger than Boris. He tried to ignore the hazing and stared at them. When they swooped too close he clicked his beak and fluffed up to look larger. Finally he moved to where thick branches protected him from the rear and resumed his ardent calling. By evening the little owl called continuously.

Later Gil and I were working inside the cabin, listening to Boris ringing his bell.

"Hey! I hear two owls calling." Gil was jubilant. We rushed out as two boreal owls flew over the cabin.

"She's beautiful," I remarked.

"Yes." Gil smiled happily. "Let's call her Bory Alice."

Both little owls were gentle and trusting. We could approach within arm's reach without disturbing them. Although tempted to stroke their feathers, we refrained from touching them. We were content to speak softly as they looked at us wide-eyed.

All our prayers were answered when they moved into the nearest birdbox. Boris's perseverance had paid off. He had won himself a lovely mate.

From December to April Boris, the boreal owl, called for a mate.

Bory Alice made her home with Boris in the nest box.

The weather moderated in late April and softened the snow. Overnight temperatures dipped and formed a crust on our snowshoe trails. We started our daily hikes on trails like paved paths, carrying our snowshoes under our arms. What fun to walk on the firm surface! We could move so much more easily and faster than slogging on snowshoes.

"I feel so light-footed." With a springy step, I skipped behind Gil, whose pack was heavily loaded with camera gear. By afternoon the sun softened the trail and we were back on our snowshoes.

Then one day in late April, as we wandered up the big river, we discovered our first grizzly track of the year. It was in the vicinity of the bear tracks in the snow which we were surprised to find last November.

"Could it be that same bear? The tracks come from a point near where we thought the grizzly could have denned." Gil pointed up to the narrow opening of a cave.

We clambered up a steep, rocky slope to the cave and searched for tracks, but the snow had melted there and we could not determine where the bear had spent the winter.

•

On April 29 we could hear Boris calling to his mate when we were in the cabin. Her muffled reply came from inside the birdhouse. Then Boris seemed to call in a loud and strange way.

'That's not Boris. It's another owl." I hurried out. A larger owl, a hawk owl, perched high in a cotton-wood near the bear trail, cocked his head and looked down at me.

"Sssssssssssss. . .sst."

Boris answered hesitantly with his bell-like notes, as if to say, "This territory is taken." He was perched behind the cabin close to his birdhouse.

Bory Alice looked out of the opening then, evidently reassured that Boris had the situation well in hand, she disappeared inside again.

Later we listened to a vocal duet by the two species of owl. First came the hawk owl's warning hisses and bell-like calls, then Boris's reply of xylophone notes.

The following day, April 30, the hawk owl was in the vicinity with his mate, but Boris and Bory Alice had vanished.

Gil could mimic Boris's call. We walked around in the woods, calling for the boreals. We wandered in all directions, even as far as a mile upstream where we had once seen Boris, and whistled their call. But the tame little owls were nowhere to be seen. Had they been frightened out of their territory? Or had they fallen prey to the larger owls? How we missed them!

The hawk owls nearby, also tame, stayed around for a while, then they, too, seemed to lose interest in the area.

Hawk owl couples seem to occupy territories several miles apart. One pair lived a couple of miles upstream and another by the lower beaver dam. Gil and I made plans to search both territories. If we could find a hawk owl nest that would compensate somewhat for the loss of the boreals. Gil might have baby owls to photograph.

A hawk owl perched high in a cottonwood. It was bad news for the boreals.

9 · Caribou Games

On May 1 Gil looked in the cabin about two o'clock.

"Come on out and bring your binoculars. I've been watching an interesting game between 10 caribou and a wolverine." I followed him out to the banks of North Creek where he sat down and focused his binoculars on the south slope of Ike's peak.

The caribou had spent months on the sloping bench between the upper and lower banks of cliffs, pawing their craters in the snow and feeding, living peaceably there, safe on their balcony of the sky-scraping mountain. Gil had noticed some of the herd bunched together while two or three individuals were running around in circles. With his binoculars he saw that the caribou were being pursued by a short-legged dark animal, a wolverine.

The caribou easily out-distanced the hard-running little carnivore. The rest of the deer watched and waited. Soon the few being chased circled around to join the herd, and the small black fellow worked hard to follow. The herd ran uphill behind a little knob, while the pursuer sometimes took a shortcut and bounded to the top of the hillock while the caribou waited for him on the opposite side. On top, Gulo, the wolverine, didn't stop, but loped down the snowy knob, in hot pursuit of the herd, which easily trotted downhill to the brow of the cliff, then split into two groups. Without a moment's hesitation, the carnivore took after one group, which went around the little hill again, while the others rested. The wolverine never stopped galloping at top speed on his short squat legs. The caribou had plenty of chances to rest. Gulo was hungry. All that meat was available on the hoof, and he couldn't catch any!

The band of caribou stayed within about a five-acre area, which is where they had been for months. They did not want to leave where it was easy going. Should they have gone out where the snow had a breakable crust the wolverine would have had the advantage. When the herd divided Gulo sometimes paused to decide which section to follow, began chasing and then doubled back to start after the other group. Now and then the

Caribou in small bands moved through woods in a general migration north.

caribou waited for their pursuer to catch up. Frequently Gulo, not as high as their knees, was right in among the caribous' legs. They had no trouble sprinting away from the little fellow. He was left far behind as they galloped to the edge of their balcony to join the others, then the herd split again, and the routine was repeated. After the caribou had rested awhile the other group seemed eager to get back into the sport. All caribou participating in the game seemed to be yearlings.

We had to admire the determination and endurance of Gulo. He tried so hard for so long.

Gil and I tired of holding our binoculars up and began checking on the strange theatrics on the mountain every few minutes. It was always the same procedure, with the persistent pursuer never resting.

"This would make a good Disney movie," Gil laughed. He slapped his hands on his knees with a galloping-horse rhythm and improvised "Lone Ranger" music, as the wolverine came bounding from behind the knob, continuing to chase caribou.

We watched for an hour and a half before the wolverine gave up and disappeared over the opposite side of the hill. The caribou watched him go, then stood looking after him, as if they wanted to play some more. Later we saw them feeding and resting.

Gil and I wondered how long the game was going on before he noticed. We had to see it to believe it. It was like a children's game of Ring-around-the-Rosie and King-of-the-Hill.

For the wolverine this had been serious business, but the caribou weren't worried. In fact, they seemed to enjoy the game. They seemed as playful with the little wolverine as when we had observed them cavorting with each other.

•

In early May we hiked several miles up the frozen North Creek and heard a new sound — water! Running water! A small stream trickled across our winter trail. Soon it was too deep to wade in our shoepacks.

"It's going to be more difficult traveling from now on. It won't be a smooth highway," Gil reminded me. We stood watching the stream, then retreated and climbed up on the cliffs to get a better view. More water was flowing every minute. Returning downstream, gurgling rivulets gushed out to join the small river, which flowed over an ice-coated streambed. All the water disappeared under the thick ice on the big river.

May 4. *16° above zero. Cloudy. Few flakes of snow drifting down. Snickaree on feeding post. Expressive tail held over her back. Lightning-fast movements. Keeps watch for hawks while feeding. We are anxiously waiting for more migrant birds to arrive.*

Gil was cooking our quota of rolled oats for breakfast, while I sat up in bed, writing.

"What was that noise?" I asked.

"That was my stomach growling."

I heard the noise again. "You've got to be kidding!"

"It's nobody else's. Nobody else here. My belly button is rubbing my backbone."

"Really?"

"I can hardly wait to get my teeth into one of my mother's good dinners. What I couldn't do to a German chocolate cake!"

We often talk about the food we will eat when back in civilization. Our moose meat will soon be gone. Then we will start using the tinned meat we have left.

Weather was dreary today. Windy. Restricted visibility. Heard a few snow buntings overhead in the snowy murk. Hope their compasses are working.

Before retiring to bed that evening, we walked out to the North Creek's gravel bar and observed for the first time the wolverine carrying a leg bone from our scrap heap. He took it upriver and into the woods.

•

May 5. It snowed all day. "It may be May time in the Arctic, but it looks like the fifth of January," Gil commented. Five inches of new snow made our house look like a gingerbread house with white frosting.

Jays were raising a ruckus north of the cabin. I went out to see if they had discovered a boreal owl.

"It's a hawk owl." Gil was there ahead of me. "I'm waiting to see where the jays come from and where they go." He was hoping again to find their nest site.

Hawk owls seemed to take possession of the area that day, spending much time perching in the Portal Trees. The jays were wild with excitement,

not liking it one bit. They chattered and clucked at the passive owls. Usually one jay heckled the owls, while the other waited its turn. When they tired of the hazing, they just vanished.

"I don't know where those jays come from. Just out of the woodwork." Gil had been out jay-watching. Again, he had been outwitted by the secretive jays.

When the owls flew across the river and called from a spruce tip on the opposite bank Snickaree dared to come to the feeding post. Approaching nervously, she ran from tree trunk to tree trunk then flew up the suet post. She was unusually fidgety, every nerve as tight as a violin string.

"I know why she's so nervous." Gil watched from the window. "Wet fur around her nipples. She's nursing her young."

Snickaree gnawed wildly to loosen a piece of tough fat, jerked it loose, then she held motionless, watching, ready for flight. Suddenly she sped down the post and over to the safety of the woods. Whenever the jays harassed the hawk owls, Snickarree added her two-bits' worth.

In the late afternoon a small gray bird flitted down to the small spot of exposed earth under the pole table. A slate-colored junco had arrived. The chickadees noticed and fluttered excitedly around the newcomer cheeping. The junco found something to eat on the bare ground, then flew up into thick branches of one of the Portal Trees to roost.

•

On the morning of May 6, caribou in small bands were moving through the woods in a general migration north. With his binoculars Gil noted that the deer on Ike were still there resting.

Later returning to the cabin for lunch after a hike to the lower beaver dam, we were weak from hunger. Eight ravens circled overhead, croaking.

"They are patiently waiting for our demise," Gil remarked. "I'm woozy with hunger."

After supper we strolled down the Swede Trail to the North Creek gravel bar.

May 6. *"Get back. Caribou!" Gil cautioned.*

A small band of caribou stood where the two rivers meet as if debating which way to go. We waited at the opening in the woods, holding very still, and saw eight caribou slowly walking north in single file. Six had antlers in velvet, just stubs. Their thin bodies and faded, raggedy-looking coats bore testimony to the rigors of winter. Lumps on their hide, where botfly larvae had developed, showed through their fur.

These are the caribou of Mount Ike which we have observed since the sun returned in February. We have evidence that they have been here since the big migration last fall.

We see the trail in the snow, where they came down today, from their winter home on the south face of Ike.

The following morning Gil stepped out to read the thermometer — a warm 34°.

"Guess what I just heard. Boris. I thought we had lost him for good."

We were happy to know he was still alive. Now that the hawk owls seemed to have lost interest in the area it should be safe for the boreals to return. We found the little owl hiding under some thick spruce branches north of the cabin. But Bory Alice was not with him. Again we heard Boris ringing his bell. He called at about one-hour intervals.

We strapped our snowshoes to our packs and hiked downstream for about a mile on firm snow to where a tiny stream began trickling into one of the sloughs, then grew until it was a murmuring brook. A new spring sound was crystalline water lapping against the ice. We saw where ravens had bathed in the shallow water and preened on the snowy bank. We could hear Boris calling from our cabin one mile away. When the warm sun had softened the snow we buckled our snowshoes on our shoepacks and waddled in the soft slush like two clumsy ducks to a pool, then we sat on the bank to watch and listen for new arrivals. We looked at the icy fangs of the Arrigetch Peaks in the distance. A buzzing fly lit on my knee then flew away. The only sound was the gurgling of the rivulet at our feet, rippling under an ice bridge, which lay astride it. Water was dripping, plinking melodiously from the icy ceiling. We heard a tinkling bird song accompanying the water music. I knelt down and peered under the ice bridge. A plain gray bird was in the frozen cavern, perched on the icy bank, singing the tinkling notes, like a little music box.

"It's the ouzel. How delightful to hear his song!"

"Glad he made it through the winter." Gil looked happy.

"It's the ouzel. How delightful to hear his song."

Hungry as bears, we sloshed toward home and heard a full-throated, finchlike warble. It came from a pine grosbeak, dressed in rosy-red feathers. What a treat to see the bright color!

Boris was still plunking his xylophone when we reached the cabin. After supper Gil called me out to listen to eerie, wild cries. From about a mile south, we heard strange loud "Eeeyou, eeeyou, eeeyou" calls, accent on the "you." It was without pause. What was it? Then there arose an answering "Yeow, yeow, yeow" for a minute or two, from the direction of South Creek. A lynx was yeowling in reply. The weird loud crying was a wailing "Eeeyou" repeated on and on. It sounded like a human child or a young bear cub in trouble. Was it caught on something and couldn't get free? We felt as if we wanted to help the poor creature. But the slushy snow would be terribly difficult to travel on now.

The cries were heart-rending. The distress calls kept on and on. Finally we couldn't bear to listen without trying to help.

We buckled on our snowshoes and slogged on south toward the cries. Our snowshoes dragged, feeling heavy as lead. The wails seemed to come from the woods south of our "airstrip," always from the same place. The crying was so loud and pathetic; surely the poor beast was crying for help. What else? We snowshoed as fast as we could but it was inexorably slow. We had to stay on the trail. Off the track we sank deep in the mush, like traveling in pudding. We dragged on in the horrible slush.

A new spring sound was crystalline water lapping against the ice.

97

The spruce grouse cock,
red eye-comb fluffed out,
put on a show near
Boris's nest box.

As we neared the woods the pitiful wailing ceased. We trudged on, expecting to hear it again. All was quiet. Not a sound from the creature. Had it moved? Was it afraid of us? We didn't know what to do, where to look.

Baffled, we waited awhile wondering what to do. We wanted so much to help. Finally we staggered back to the cabin dog-tired. (Much later, back in Oregon, we heard these cries from a female porcupine in heat.)

That same day, Boris called at about one-hour intervals throughout the day. In the evening he called continuously. In bed, we listened drowsily as the little owl rang his bell. It was our lullaby.

About midnight we awoke with Boris calling, calling, calling. There was no darkness this time of year. Gil went out in his pajamas and saw the elfin face of a boreal looking out of the nest box, calling.

"Now, you can be happy. Boris is back in his nest box," Gil assured me as he climbed back into our bunk.

"Uh-huh!" I agreed sleepily.

Sometime during the night a bear passed by the cabin and a moose approached within 40 feet as we slept.

"Wish we could have seen them," Gil said as he studied their tracks next morning. "If we didn't have to sleep, we could see more."

•

May 8. The warm sun triggered avalanches that thundered down the south slopes of the mountains. What a contrast to the silence of winter!

Gil had fun photographing a very tame spruce grouse while she foraged near the bear trail, plucking into the vegetation, eating alder buds and probably finding some of last year's bearberries. While we worked around our house she stepped sedately across our dooryard like a tame chicken. Later, we found out that the spruce hen had a mate. He was putting on a show near Boris's birdbox, with brilliant red eye-comb all fluffed out. Like a turkey gobbler, his tail edged with chestnut brown was spread like a fan. He was quite a spectacular creature. He strutted majestically over the snow to bare ground. His fanned tail swayed from side to side as he ruffled up his neck feathers. He rattled dry leaves with his beak and uttered low crowing notes while performing for his mate. All the while, a few feet away under a spruce, his lady sat with half-closed eyes, apparently unmoved.

The plain little hen roosted for the night about six feet up in a spruce, near Boris's birdbox.

Gil had fun photographing a very tame spruce grouse hen while she foraged near the bear trail.

10·Birds, Birds, and More Birds

 May 11. Migrant birds swarmed in. Boris was calling when we awoke. Hearing the mew gull and varied thrush, we tumbled out of bed and rushed out just in time to see a flock of common redpolls alight in the willows and spruce in our yard. One of the males, perched nearby in the sun, was a real beauty with scarlet marks on head and breast.

Two mew gulls, cruising at treetop level over the big river, heading downstream, met a silent Canada goose winging north. My heart went out to him, as I thought of Aldo Leopold's statement in *A Sand County Almanac* about lone geese in the spring. "They are bereaved survivors of the winter's shooting, searching in vain for their kin." Along with Leopold, I "grieved with and for the lone honker."

We turned to hear the drawn-out quavering notes of a varied thrush, first high-pitched, then an octave or so lower, with long pauses between notes. It perched on top of the tallest spruce, like a preacher, turning to call in every direction.

Snickaree was scolding from atop a birdbox at the Swede Trail. We found she was denouncing a hawk owl perched close to her nest. How the owl got there without the jays discovering it is a mystery.

Three robins flying north met a single marsh hawk going downriver, winging just over the trees. A whirring of many wings overhead proved to be fast-flying ptarmigan. Across from the spring pool, perched in the willows of the island, a tree sparrow piped his poignant song.

May 11. *We are surrounded by slushy six- to eight-inch-deep snow. Water stands in every pothole and gully. The rivers are slowly rising. Hope the water doesn't spill through from the slough to a pool downstream, where Gil has spotted some grayling he hopes to catch.*

Most of the ground is still snow-covered and snowshoe hares, grayish-brown now, scurry quickly over the snow, staying mostly on bare ground under the bushes. They are edgy and nervous, seeming to realize their cryptic coloration is out of whack.

The arctic terns were catching sculpins in the riffles over the river bars.

But our dooryard is free of snow. We step out of our cabin on bare ground. It is so good to walk on the soft good earth again. We often sit on a bench made of half a log, soaking up the sunshine, on the south side of the house.

May 12 was another exciting day. Afraid we might miss something, we hurried out to see what was happening. How wonderful to witness the phenomenon of the arriving flocks! We were awe-struck by the magnitude of the movement. Suddenly there were birds everywhere. Their excitement was contagious. As different species made their appearance, we marveled at the endurance and instinct which guided their journeys north. We wondered how they had fared. Because of our winter solitude the impact of this bustling activity was enormous.

"Just stop and think about it," I mentioned to Gil. "This wondrous event is happening all over the Northern Hemisphere."

Around our yard varied thrush, pine grosbeaks and juncos were singing. There was a snowshoe hare at the edge of the willows on the island. Snowshoe Harry and a friend were in the woods by the Swede Trail, hopping leisurely, stopping to nibble twigs of willow and alder. Jays were secretive and unobtrusive as they pecked at a bone on the pole table.

"Hear that whistling?" Gil tipped his head. "It's a snipe winnowing (making his wings whistle)." We both saw it circling in the blue sky, up and down in a "bent wheel" manner.

"Teedle-lee Teedle-lee Teedle-lee." It was a yellowleg's call.

"I'll never forget that darn bird," Gil remarked. Once when we searched for a nest in a yellowleg's territory the incessant shrill cries of the shore bird got on our nerves so much, we gave up in disgust and left the area.

Gil went inside to fix lunch while I kept a vigil outdoors. We were afraid we'd miss something. I sat on a bench near the cabin when, through the trees, I saw grayish-white fur coats filing past on North Creek. I tapped the window and told Gil. He hurried out with his camera pack, arriving in time to watch six caribou approach in single file on top of the snow crust. Three bulls and three cows walked out on the ice of the big river and stood in a huddle, as if trying to decide which way to go. Leaderless and hesitant, two started up the big river, two started toward South Creek, and two in another direction. The pair heading up the big river spooked and ran back. They were disorganized and nervous. Finally all the animals fled up in the direction of South Creek.

Gil took his fishing pole and we hiked down toward the pool where he had seen fish. He hoped to catch some grayling while the water was still clear. We saw our first swallow. Its twittering call identified it as a violet green. We prowled through some short willows and found a small pond in a marshy area. There we identified our first American widgeon in our travels to the Brooks Range. We were surprised to find a pair of the ducks holding their heads down, trying to hide. A

The snowshoe hare changed color as warm weather approached.

pair of pintails were with them and three yellowlegs probed around the grassy edge of the small pond.

After settling down at the pool, Gil began catching grayling. An ouzel watched him awhile before she went about her own business swimming underwater, searching for fish. She picked insects off the snow of the bank, then swam like a tiny duck on the surface of the water, making a small wake. Then she flew upstream and back again, always busy, busy.

Nearby a marsh hawk, in courtship display, was putting on a spectacular show of roller coaster swoops and barrel rolls.

A flock of 20 pintails flew by heading upstream. A lone, sparrow-size bird flew over heading north, calling "Pip-it — pip-it" all the time while flying. It was a water pipit. Later on a flock of 15 or 20 pipits followed.

"Here come two ducks from the north," Gil called. I crouched low as two mallards whizzed by close over my head. A lone pintail arrived. All the while mew gulls patrolled up and down the stream.

"This is a bird watcher's paradise," I commented.

Gil caught 24 fish before we headed home. My fisherman had finished just in time before the rivers began spilling over into the slough, roiling the water.

Hiking home, carrying our load of fish, we anticipated a feast.

"Kee-kee-kee-kee-kee." Shrill, high-pitched cries greeted us from the area of Boris's nest box.

"I'll be darned. Sparrow hawks!" Gil announced. A female kestrel sat on top of the birdhouse.

"Oh, no! Not Boris's nest box!" I wailed. But the hawks stayed around, obviously interested. Poor Boris.

"Too bad," Gil said. "But Boris fooled around too long."

Late in the evening we heard a familiar song.

"I sing in a minor key!" seemed the words to go with the sweet clear notes whistled by a Gambel's white-crowned sparrow.

The following morning Gil got up and announced that the kestrels were wasting no time.

May 13. *The sparrow hawks call "Kee-kee-kee-kee-kee" sharp and clear. Tercel (the male hawk) displays with roller coastering and barrel rolls. They make many trips into the birdbox. Lots of aerial acrobatics. Seem to be having a great time! More going in and out of the birdbox. Sometimes both are inside at the same time. Then more roller coasters and showing off. In p.m., we saw them mate several times on tip of a spruce. He continues his exhibition, with all kinds of wild diving around, ending up with nose-diving directly into the nest box. The demonstrative little falcons are beautiful and interesting to watch.*

In the afternoon we found a jay was doing an excellent imitation of the hawk's call. Things are confusing enough, without the jays complicating matters. Later they were mimicking the mew gulls, the rascals.

Boris sat disconsolately on a lower branch of a spruce near the birdbox, calling softly.

Kestrels settled in Boris's birdbox. They made many trips into the box.

I went out to visit him, standing very close. I spoke gently to him. Poor Boris. I was sorry he was having such a hard time.

Many interesting things happened on this day. Three snowshoe hares hopped around in the yard, blending in with the brown colors. Twice I nearly threw the dishwater on Snowshoe Harry before I saw him.

A fox sparrow chanted his bouncy song, a pair of Bohemian waxwings stopped by and later a flock of 15 or so flew over. A snipe winnowed overhead, while four mew gulls flew back and forth all day. We heard the first full rollicking spring song from several robins.

Adding an ominous note, a sharp-shinned hawk perched near the stream, silently watching for prey, while a female marsh hawk winged past. Several flocks of ducks whizzed by, flying low and fast, and were gone too soon to be identified. We did identify 16 pintails and 2 widgeons and Gil saw another lone Canada goose.

We wish we could have identified all the birds going past, but that was impossible.

"Here comes the water!" Gil was looking out the window. Muddy water was flowing past our cabin. We heard bumping noises from the big river. We ran out to see what was happening.

The great moment had arrived. On May 13, breakup came on the big river. Great pans of ice, a foot or more thick, were jammed, twisted and stacked up in a chaotic heap. Muddy water was roiling, churning, and bringing more ice and debris down to the ice jam. Upstream immense blocks of ice crunched forward, carrying drift logs. The river was jammed for half a mile upriver. Scary to see an ice jam up there. What power was in that water! It could cut loose and flood where we stood.

Gil and I had never witnessed a breakup before. We stood transfixed. Finally we pussyfooted out on the ice floe for the feel of it. The wild water thundered and roared all around. Churning muddy water, carrying chunks of ice, was rushing in all the streams now.

"The cabin! The flood could wipe it out!" We raced back to check on our house. Water and debris surged in the channel 15 feet from our cabin. We watched anxiously as the flood waters

May 13,
breakup came on the big river.
Great pans of ice jammed up.

rose higher and higher, close to overflowing the banks. But the water stayed within the channel and by evening we saw it was lowering slightly. The crest had passed.

"We're isolated now," Gil observed. "We won't be able to cross the rivers for a long time."

"Well, there are lots of places we can visit. We can still go up the mountain and explore all along this side of the river."

That evening Gil climbed the crude ladder to the meat cache and brought down the last hunks of moose meat for cooking. A jay was stealing meat scraps tossed out on the river bar for the gulls and, adding insult to injury, mimicked the mews' calls.

Our new neighbors, the kestrels, perched near the birdbox and were quiet. Boris also was quiet. With the loss of the nest site, perhaps he had given up.

May 14. Following the riverbank, we headed down toward the lower beaver dam. On the sunny south side of a little hill, I walked ahead of Gil. At the edge of a snowbank, we were delighted to find the first flowers of the season. They were purple pasque flowers announcing spring. More beautiful than we had ever seen. All the more welcome, because most everywhere we looked snow covered the ground, except for a few small spots of exposed earth and rock.

"Let's check on the beavers. I sure hope they made it through the winter." I had worried lest the floods rip out their beautiful dam spanning the small channel. Rounding a bend, Gil stopped and raised his glasses.

"If they're out, let's not scare them," he cautioned. "Oh, now I see them, in front of their lodge."

Two amphibious rodents were obscured by willow brush held fast in the ice, the remains of their last winter's food cache.

"See," Gil pointed out, "they've gnawed a hole in the ice in front of their lodge, or they'd still be imprisoned. What powerful teeth and jaws they have."

The two beavers were out of sight when we approached quietly and sat down to watch. One appeared at the hole and climbed out on the ice. Soon a smaller one came out to join its partner. They touched noses. This was the same couple we watched last fall. They nipped off some frozen-in willow sticks and sat on their haunches, chewing the bark with evident enjoyment.

On the way back home we passed the water ouzel and noticed that there were now two of them.

"They'll be moving up into the mountains soon." I wanted them to remain with us. "Probably they will nest near a waterfall. They seem to require water music."

Near a marsh we passed a moose browsing on short willows. We noticed she was heavy with calf. She swung her great head around to watch us go by, then resumed browsing.

Small groups of caribou came by traveling north.

Purple pasque flowers announced the arrival of spring.

Over the big river an arctic tern flew with slow and easy elegance upstream, then back down, stopping to hover over riffles of the slough. Each year our first sighting is a delight. Such grace and beauty in their languid wingbeats!

May 21. *"What fun we are going to have at Wildhaven," Gil said, as he brought me my morning cup of coffee. (Wildhaven, our wildlife sanctuary, is our home in Oregon.) "We'll have fun eating a big stack of hot cakes, all we can eat, and lots of rice, and rich cakes with whipped cream."*
"We'll have fun getting fat," I warned. "But now, please bring me a couple of hot sweet rolls with lots of butter." We dream of all the good food we'll eat in civilization.
"I've been hungry since last February," said Gil.

Summer weather had arrived. Temperatures were in the 60s and birds were active everywhere. The arctic terns were catching sculpins down off the river bars. Hovering to spot the tiny fish, they seemed the most beautiful of arctic birds, graceful as ballerinas, with silvery mantles, snow-white wing linings, crimson beaks and feet. When a fish was caught the tern announced it with low hoarse calls and presented it to his mate.

Hiking in the Enchanted Forest we heard "Pik—pik." Mrs. Three-toes was industriously excavating a nest hole in a dead spruce, three feet above ground. Sawdust was strewn on the ground below. The woodpeckers were secretive now. There was no more drumming.

We passed a moose
browsing in short willows.
She was heavy with calf.

106

Not far away, on the bank of North Creek, the goshawk sat motionless on the stick platform, utterly silent. The new branches added this year formed a banister around the edge, helping to screen the bird from our view.

Nearby North Creek was roiling and hurrying toward the big river.

Next day we had a lunch of diluted milk and the last of our moose meat, which was mostly gristle, not very tasty eating but at least the gristle filled our stomachs.

I sat outside the cabin on a bench writing and watching two snowshoe hares playing tag at the edge of the willows on the island across the spring pool. One chased the other into the open, then it skirted the willows and ran back in, with the pursuer about 10 feet behind. Still out on the bar, the latter stood on its hind legs to see where its playmate had gone, then he doubled back and ran in to intercept. A moment later I saw them zipping across, just inside the brush.

In the afternoon, a pair of jays brought their fledglings to our yard. Their tails were adult length, but they still wore prominent yellow flanges at the corners of their mouths, typical of all that type of baby birds.

"Those darn jays sure kept their nest site a secret." Gil voiced his frustration. "Now I suppose they expect us to help feed their kids. Well, we're about out of food."

I chuckled at how they had outwitted us.

The next morning, Gil summoned me when a porcupine waddled up to the spring pool near the cabin. We went out and witnessed behavior similar to that described by Sally Carrighar in *Wild Heritage,* of a female porcupine hankering for a mate.

"Let me see you do your trick with this one," Gil challenged.

May 23. *I stopped about 40 feet from the little beast and began humming softly. Porky turned and hurried toward me, stopping about eight feet away, then stood up on her hind legs. I kept on singing while Porky watched and listened. She chattered her lower jaw, yawned and listened. Chattered again and yawned. The creature turned to go, then came back closer. Eyeing me, she chattered and yawned again. Three times she started to leave and came back to chatter and yawn.*

"Wonder if she thinks you are a male porcupine!" Gil grinned.

The following day, Gil met the porcupine in the woods. Concealed in the brush, he watched the animal, which was playing all by herself, like a child, having a mock battle with an 18-inch cottonwood sapling, backing up to it and batting her tail against it. She waddled over to a small spruce, climbed up a few feet and hugged it. Returning to the sapling, she lay over it, bending it over and riding it to the ground closing her eyes. After about three more mock battles and backing up to several more saplings to flip her tail back and forth against them, the porcupine padded away through the woods.

An arctic tern flew with slow and easy elegance.

11·Bear Raid

May 29. I hurried with the household chores so that I could join Gil out on the big river sand bars. A pair of harlequin ducks were resting on the rocks in the middle of the rapids near the "airstrip." The harlequins loved the turbulent waters, which splashed over them where they rested. Several flickers called "Kak, kak, kak, kak," in several directions from the woods.

"That's their happy call," I noted. "They're singing."

"Let's go home and start lunch," Gil urged. "I'm starved."

"Wait awhile," I coaxed. "We might miss something here if we leave now." We waited 20 minutes longer, then walked leisurely back to our slough. In no hurry, we sauntered down the gravel bars toward home. Gil was leading. Abruptly, he stopped.

"Oh-oh! We've had company!"

"Eskimos?"

"No. It's a break-in."

Then I noticed the windows looked strange. No bars across. We got closer. I ran ahead to see the damage. Gil called me to come back and get behind him.

"Our visitor may still be inside." Gil loaded the rifle. There was a clatter and then a thump inside. We saw a dark form inside through the window.

Gil and I stood on the sand bar below the cabin. A large black furry face appeared at the double window to the right.

"A large head. It's a grizzly!"

The bear tried to get out, but bars still in that window held him back. He disappeared. Suddenly the bear was at the other window, one with broken bars. We saw his light brown snout and knew it was a black bear. He scrambled out in no time.

"Don't shoot him," I begged. Gil was taking aim.

"I'll shoot just over his head to scare him." The bear galloped to the back of the cabin. Gil fired over his head with a fearful blast that shattered the serenity of a whole winter and spring in the Arctic. The creature disappeared in the brush.

"That's a big black bear."

Viv saw the bear in the yard about 20 feet away. She screamed for Gil.

The window frames were reduced to kindling. Tatters of cloth lay on the ground outside.

"Look at my curtains," I wailed. Tatters of green terry cloth, trimmed with red rickrack, lay on the ground outside. The window frames were reduced to kindling. We had worked so hard to make the boards for them.

"That bear will be back," Gil predicted. We scrambled up the bank.

"I shudder to think what the inside of the cabin looks like." Fearfully, I opened the door. First I looked for the food tins. They were all in a row under the counter, just as we had left them. But broken glass was everywhere, on the counter, all over the floor. Every window pane, except one which was out, was shattered. Bruin struck the panes such smashing blows as to send splintered glass against the opposite wall. The curtains were in shreds. He had entered the cabin over our eating table.

The beast had not been in the cabin long, but nevertheless had slurped up Gil's sourdough starter, finished off our last two cups of rendered moose fat and had just started on a two-pound can of ham when we interrupted him.

"We are two lucky people! He didn't have time to clean out our food supply."

"And he could have torn everything to pieces, rubber raft and all."

"Hey! How about my exposed film?" Gil yelled, as he ran behind the cabin. "Thank God! He walked right past it." Gil brought in a five-gallon can nearly full of his year's efforts at photography. He patted the can fondly. "Whew! I'm putting this up on the cache . . . now!" He went out to carry it up the ladder. The bear had bitten through another can lying beside his treasured exposed film.

"Thank God he didn't come yesterday." We had been gone all day looking for new birds.

"Yes. We'd have been hungrier than we have been."

"He'll be back tonight," Gil predicted.

"Our beautiful windows!" I started to pick up the broken glass.

"I'd say we're lucky. We've had a good lesson. Everything we value has to be put up out of his reach."

Still picking up broken glass, I was distracted when I heard a flicker calling from the bear trail. Gil had mounted a nest box out there especially for Snickaree.

"Do you think a flicker will take over the nest box we put up for the red squirrel? I'm going out to check." I walked out to the Swede Trail.

"Yikes! Here's the bear!" I yelled to Gil. "On our Swede Trail!" I raced back to the cabin. The beast crashed away through the underbrush toward North Creek.

Gil had not expected him back so soon. "We'll have to shoot that bear."

"No," I pleaded. "He doesn't know he's doing anything wrong."

"He'll keep coming back. He's not afraid." We hadn't long to wait. Gil went out to North Creek to check and see if the bear was around. I hurried to carry some things up on the cache. Back in the cabin, I kept the door open. Glancing out, I saw the bear in the yard, about 20 feet away, coming for the open door. I screamed at the bear. Loud! I hoped Gil could hear. The beast hesitated, then kept coming. I screamed again and slammed the door. I knew how dangerous bears could be and stood trembling inside the cabin.

When the beast saw Gil approaching it turned and waddled into the woods. Five times the bear came back. In between times it skulked in the underbrush around the cabin.

"I'll have to do him in," Gil announced. "I hate to do it, but it's dangerous to have that animal around all the time. He has no fear."

At times Bruin came padding down the bear trail or appeared from the thick woods. We worked to clean up the mess in the cabin and began hauling our food up the ladder for safe storage.

"Nothing is safe in the cabin. We don't dare stop guarding it."

Late that afternoon I stepped out to empty the dishwater, turned to go behind the cabin and met the bear, eyeball-to-eyeball. I screamed, dashed back in and slammed the door.

"With all that screaming you're going to make a nervous wreck of that bear," Gil commented dryly.

"I'm the one who's going to be a nervous wreck," I panted.

Bruin was undaunted. He'd had a taste of moose fat and Gil's sourdough. He knew a good thing when he tasted it. We heard him opening a can of washed fireweed, which sat in the shade near the cabin.

"There goes the fireweed for our salad!" We heard him about five feet from the door.

I was sorry for the bear. This was his land, not ours. I didn't want to destroy him. Yet he was dangerous. We could bump into him at any time.

"Couldn't we teach him a lesson? Maybe you can pepper him with bird shot," I offered. Gil had brought that for an emergency, in case we lost our food and were starving.

"I don't want to wound him either."

"He won't learn until he's hurt." I pleaded. "Please don't murder him."

"I hate to make him suffer, too," Gil said. He heard the big fellow snuffling around the back side of the cabin.

"All right," Gil decided, "we'll try to teach him a lesson. If it doesn't work I'll have to kill him. I hate to do that though." Gil took the shotgun and, as an afterthought, also picked up the rifle. "I'll use this only if necessary. If he moves away I'll give him a dose of bird shot."

Gil opened the door and walked straight away from the cabin and looked around the side of the house. There was the bear. Bruin started ambling toward the woods. We both yelled and he went about 60 feet, then Gil blasted the animal in the rear with bird shot. It spun around, then rocketed down the trail into the forest.

"I just hate to think of that bear suffering." Gil was remorseful. I lay on the bed crying. Then I did a double-take. "Bears have tough hide, don't they?" My spirits lifted at the thought.

"I've heard that." Gil smiled with relief and we both felt better. We hoped Bruin had learned to avoid man and his cabins from now on.

To calm our nerves we decided on the luxury of a cup of coffee.

"What a terrible thing it is to wound an animal." My naturalist sounded sad. "That bear didn't let out a whimper when it got blasted. I'm glad I didn't kill him."

"He's cute," I added. We kept watching the bushes, but we really didn't expect the bear to return.

It took many trips up the ladder to move our supplies up on the cache. First thing up was the rubber raft, which was our contact with the outside world. Then we hauled up the Klepper tent, air mattress and all our remaining food, except what we needed for the day.

"We had felt so safe. The beast could have come inside the cabin while we slept." I counted our blessings. Gil's exposed film, all our records, my diary, our food — all were safe.

We put new bars across the window and fastened plastic over the openings.

"Now if a bear tries to get in he would make enough noise to wake me," Gil promised. We slept soundly.

Next morning, I woke with a start. I heard a low growl very close.

"That's my stomach growling," Gil informed me.

•

Suddenly we were walking in springtime. Snow was gone in the valley and there was a refreshing new greenness over the land.

Shrubs and small trees, with emerging new leaves, formed a frothy green veil under a canopy of bigger trees in the Enchanted Forest. Green crept up the mountain slopes.

Flowers sprang into bloom everywhere. Rosy-red arctic fireweed carpeted the gravel bars and sweet vetch bloomed in bright pink profusion, filling the air with heavenly perfume. It is considered a poisonous plant. In the early days of arctic exploration Sir John Richardson and his men mistook it for the edible Eskimo potato and reported that all the men who ate it became ill.

Anemones, five-petaled white stars, studded the chartreuse sphagnum moss on the forest floor, along with shin-high Lapland rosebay and Labrador tea.

"Kee-kee-kee-kee." Home from the hunt, the male kestrel arrived to perch near the birdbox with his trophy, a redback vole. He continued his shrill calls until his mate looked out of the nest hole. Presently she came out and settled beside him.

"Look how greedily she snatched that food." I was appalled at how she jerked it away from him. "He works so hard to feed her. Couldn't she be more gracious?" We watched her tear at the meat and gulp it down.

"She's just showing her dominance," my husband smiled. "Female hawks are larger than the males and dominant. That's natural. Let's just call it a matriarchal arrangement."

Undaunted the male hawk flew against a strong head wind over the big river and began soaring in circles, gaining elevation, riding an invisible spiral escalator. The small falcon went around and

Sweet vetch bloomed in pink profusion, filling the air with perfume.

around without a wingbeat, riding the thermals until he was a mere speck in the sky. Gil and I watched through binoculars as the intrepid hunter climbed until he was higher than Noon Peak. He sailed over the summit on rigid wings and vanished into a gray cloud. A couple of hours later he came back with another vole to offer his partner.

Around the first of June we found fresh tracks of a newborn moose calf traveling with his mother. They passed close to the cabin without our seeing them.

"Perhaps that was Moosette." I wished we could have seen her calf. Next day, prowling through brush on the island south of the "airstrip," we found a reddish-brown, newborn moose calf, lying dead in the woods, curled up without a mark on it.

"It's been dead about a week," Gil surmised. "Looks like it died the same day it was born."

Snowshoe hares now wore summer coats of brown, while their oversize hind feet remained white. They were not so nervous now that they were camouflaged to blend in with browns of the earth. Baby rabbits, born with this concealing coloration, held motionless, trusting that we did not see them. Usually we didn't. They were difficult for us to spot when walking in the woods.

Gil and I became obsessed with bird-watching, adding more and more species to our list. A favorite spot for observing incoming birds was out on the wide river sand bars. With a beautiful double-noted whistle a black-bellied plover, in striking black, white and mottled plumage, alighted near the edge of the water. It was accompanied by a tiny sandpiper. They probed the mud and fed side by side like a domestic hen and her baby chick. The semipalmated sandpiper moved away as they probed busily. When the large plover whistled his melody the little sandpiper flew over to it, then they flew away together up the big river.

"That's a strange association," I mused. "Do you suppose that little sandpiper is imprinted to the plover?"

"Could be," Gil looked puzzled. "After reading Conrad Lorenz's *King Solomon's Ring* I know that stranger things have happened."

"There's a different bird." I pointed to a low-flying sandpiper type headed straight for us.

Baby hares are born with protective coloration. They held motionless so we would not see them.

"That one's easy," Gil said as it flew past at eye level. "It has a bright cinnamon belly and white line across the breast."

"Wow! A dotterel. What a wonderful close view of a dotterel!" I threw my hat in the air as the visiting shore bird from Asia flew downstream out of sight.

"Swoosh!" Rushing wings whirred directly overhead. Startled, we looked up as a golden eagle came out of a skydive a hundred feet up. It was showing off for its mate. Both eagles soared toward Noon Peak.

Arctic terns hovered over the river riffles, poetry in motion. Nesting on a low island upstream two mew gulls cruised up and down the river.

Gil looked downstream. "There's a new bird." A bedraggled-looking snow goose stood alone on the gravel close to the water.

"It looks lonely and tired." I wondered how it got separated from the flock. It watched us warily. Gil approached and got one picture before the white goose lifted and flew upstream. Setting its wings, the bird prepared to land on the gulls' island. An incubating gull intercepted it so the goose maintained elevation, turned back and again tried to land, but was thwarted by the couple who had possession of the island. Seemingly near exhaustion, the forlorn goose kept flying low downstream.

In early June we had our first real rain of the season. How sweet the air smelled! Loving the rainy weather, many species of thrushes joined in song. The Swainson sang, "Can't you hear I'm a Swainson's thrush?" We heard his lyric song around the clock. His music was divine. And then there was the burry twang of a gray-cheeked thrush, while the waterthrush added his loud, bright melody. With the varied thrush contributing his poignant, slurred notes, along with the robin's caroling, it was a most enjoyable concert.

We went to bed that night with plans for a long hike tomorrow for more bird-watching.

"Bear! Hey! Get out of here!" Yelling, Gil fell out of bed and ran to the window. Groggy with sleep, I was hardly aware of what was going on.

It was midnight. Gil heard ripping plastic. A dark furry face looked in. Big paws clawed at the plastic. Then Gil yelled and the bear backed off. Gil looked out and saw the beast nonchalantly padding down the bear trail to the Portal Trees. Bruin stood up on his hind legs and, with his back to a tree, began to scratch.

"Come look at the bear scratching at the Bear Tree."

I remained in dreamland.

"I thought I taught that bear a lesson. Lucky I woke up."

Suddenly I was jolted awake at the thought of bear inside the cabin with us. I jumped out of bed and stumbled to the window, just in time to see the bear ambling down the bear trail. He returned to the Portal to scratch his back again, then proceeded down the Swede Trail toward North Creek. Later, we heard him crashing through some brush behind the cabin.

We waited for the animal to return. He probably

Spruce trees lined
the river near a spot
where blackbirds gathered.

would keep coming back, so we rigged up bear alarm systems at all the windows, with metal plates to fall down on empty tins.

"Our food is on the cache, but the tins around the trees supporting the platform may be too short," Gil said. "We'll have to fix them tomorrow."

"We're not safe in the cabin either. If he got inside . . . oh boy!"

We climbed back in bed to wait for the beast to come back. I couldn't sleep. I was really jumpy. I thought I could hear him padding around outside. About two hours later I was about to doze off when Gil jumped out of bed.

"There's a plane circling over the cabin." He went out in his pajamas. "He's looking for a place to land. It's a very small plane." The aircraft circled several times, came lower and we thought it was landing.

"He chickened out."

The plane headed downriver, turned again and came back to the "airstrip." We heard the loud crunch as it hit the gravel.

"He's down." Gil pulled on socks, pants and hip boots.

"You stay here in case the bear comes back. He's probably lying out in the bushes waiting. Bar the door after I leave. You have the shotgun." He disappeared down the bear trail.

On the gravel bar, Gil met Sandy Hamilton and his friend, Bob.

"You're the second visitors we've had tonight." Explaining about the bear, he brought them back to the cabin and all three men ducked low to enter.

Sky and river glowed in the soft colors of an arctic sunset.

They were our first human company for many months.

Sandy expected us to float downriver soon after breakup. Nobody had seen us come out so he was concerned about our safety. Furthermore, the Natives were worried about "those crazy gussicks up the river." We sat and talked for two hours.

Sandy and Bob, teachers at Hughes and Huslia, were concerned about the way wildlife is persecuted by some people in Alaska. They told sad tales about hunters running down wolverines on snow machines.

"Black bears are more dangerous than grizzlies," our visitors told us.

"If a black bear gets into the cabin with you — look out!" they warned. They didn't believe we could teach the bear a lesson. "He will keep coming back until he is destroyed — or you are!"

When our friends learned we were short of carbohydrates, Sandy said he had some extra oatmeal on the plane which he could give us. Gil walked out to the plane to see them off. He came back with a bag of 20 cups of oatmeal, mixed with powdered milk and raisins, 2 dehydrated dinners of chicken and rice and chili (4 servings each) and a package of pemmican from Bob. Glory be! This food would give us energy to get on with our work, hiking around to get pictures of birds and flora.

By then it was 4:00 A.M. (we had set our watch by the sun) and we'd had only two hours' sleep. We climbed back into bed again with bear alarm systems rigged at the windows. We slept about four more hours.

After our midnight visitors, Gil and I worked the entire day adding more tins to the legs of our cache, making the tin barriers twice as long as before. Then we put all our food on the platform, except supplies for breakfast. We carried up all our clothes except what we wore every day. Down jackets, leather mitts, climbing boots, snowshoes, all went up.

All day long we kept looking around for the bear, expecting him at any moment. He could be lurking in the bushes, behind the cabin, or screened by the new green leaves on the alders and willows. Every time we stepped out, we looked cautiously around the corner before venturing out away from the cabin. We rigged the bear alarm system again at bedtime.

The next day was much cooler for the hard work we planned. We needed bearproof covers for our windows before we could leave the cabin for a day's hike. Out in the Enchanted Forest we found a suitable log with which to work. Gil carried it home and we both worked at sawing it lengthwise to make planks for heavy shutters. It took the entire day of hard labor to saw and trim it so it would fit over the windows, but we couldn't leave the cabin unprotected.

June 7. *Working on the shutters, we kept close to the cabin to protect it. I took one short jaunt to North Creek. Near some bushes, I stopped short when I heard a distinct growl close by. My heart stopped. I heard another growl. Then I realized it was my stomach growling. Whew!*

12·Summer Solstice

June 8. With a little extra food to provide enough energy we made ready for a day-long hike. After fastening our new heavy window shutters in place we felt reasonably sure a bear couldn't get in the cabin.

Facing a chilly wind on the North Creek gravel bars, we headed upstream. We looked up and saw a long-tailed jaeger flying north over the river. We were pleased to see this gull-like oceanic species, with its unusually long, slender tail, familiar to us from previous summers spent studying birds on the crest of the Brooks Range. Entering the woods, we proceeded toward a rounded outcrop we had seen in the distance, at the foot of Lamb Butte. Rhododendron was blooming prettily, along with ground-hugging purple saxifrage and dryad, festive with its white "daisies." Lowbush blueberry was decorated with delicate little bells.

We enjoyed lunch with a view of two small lakes below. Phalaropes were circling there and an arctic tern hovered gracefully, first over one lake,

then over the other. We dropped down to visit the birds there.

On the first lake which we visited a northern phalarope swam around, jerking her head back and forth as chickens do when walking. We struck out through open spruce woods, carpeted with reindeer moss and blueberry bushes, heading for the second lake, passing by a few paper birch, along with the more common variety, dwarf birch. Soon we glimpsed a variety of bird which we hadn't seen before in the Arctic.

"Surprise! These are red phalaropes." Gil was happy to see these little brick-red birds, an oceanic species, our first sighting of them. Their backs were decorated with a rich, golden filigree pattern. About two dozen phalaropes were spinning in circles as they picked what appeared to be small water skaters off the glassy surface. The phalaropes swam around like tiny ducks, with jerking heads, making wakes like minute motor-boats.

"These birds are larger than the northern variety." Gil had fun taking their pictures. "They're

*On one of the little lakes
we found red phalaropes,
a rare oceanic species.*

The red phalaropes were all females, a liberated lot, whose mates were busy incubating.

so active. I wish they'd hold still." The brick-red birds were all females, taking life easy, while their dedicated plain-colored mates were incubating their eggs in nests hidden in the marsh.

"These females are truly liberated," Gil laughed, "not like female ducks, who are deserted by their mates and left to raise their young."

"There's a tern on her nest." Gil pointed to what I thought was a piece of bleached wood on a grassy islet near the opposite shore. Red phalaropes swam around the pond, making ripples on the glassy surface which mirrored the tern; she sat like a queen on a throne on her islet. Little phalaropes provided interest and company for her during the long hours of incubation.

When I waded in to take an egg count pandemonium broke loose. The terns and all their neighbors voiced their indignation as they cried and flew around the invader. Terns swooped over and gave me angry pecks on the top of my head with their sharp red beaks. Ouch! I counted two eggs.

A hundred feet from the lake I stepped over a bleached log, when there was a thump of wings on the log. An incubating yellowlegs came off her neat nest of four splotched eggs three feet from me. Her perfectly camouflaged eggs were nestled in a depression in dry reindeer moss and brown leaves. The birds' shrill cries rang in our ears as we left in a hurry to head for home.

When Viv waded in to take an egg count the terns swooped over and pecked her.

We had not gone far when we noticed a few quaking aspen mingling with the paper birch growing on the dry south slope of a low hill. Adding to our surprise, we also found common juniper, a creeping shrub, which we had not seen before in the Brooks Range, not far from the lower beaver dam.

Proceeding toward the cabin, we waded through a riot of colorful flowers. There was purple lupine, pink sweet vetch, blue larkspur, yellow arnica and shooting stars in bright rose-red, all interspersed with creamy pale pallid paintbrush.

"Quick! Three beers!" An olive-sided flycatcher called loudly, staking out a claim near the cabin.

"That's the thirsty bird," Gil laughed.

•

Through June the weather was generally drippy. Rivers nearby ran high and muddy, restricting our travels to the south and east.

June 18. I wish it would rain hard and get it over with. Our roof does not leak, and we are cozy as can be, but we long for the flood water to recede. We both are very thin. Gil laughs at how the seat of my trousers hangs down. Nothing to fill it out.

"No popo," laughs Gil. Making more holes in his belt, he cinches it tighter and tighter until there are gathers all around his waist.

In the afternoon we hiked past Ouzel Springs and found death camas and nagoonberry blossoms. We sat on the banks of the big river above the high, swirling water to absorb the view of the lush growth of spring leaves in the valley, a mossy velvet carpet at our feet.

We skimped on food that day, saving more for a climb up Ram Peak when weather improves. Our footsteps dragged walking home. We were still tired after our meager lunch of two pancakes and one cup of milk. A supper of German potato balls, creamed ham, cooked fireweed and chocolate gelatin revived us so that we could enjoy the evening. We read to each other and played a game of rummy on that dark and rainy evening. I read Gil to sleep.

•

Summer solstice, June 21. The longest day in the year, called Midsummer Day in the Scandinavian countries, was a cause for a celebration. We planned an around-the-clock climb of Ram Peak to view and photograph the midnight sun.

We set out shortly after noon with a cloud of mosquitoes following each of us as we hiked up North Creek. We stopped on a gravel bar to dope up with insect repellent, then the blood-thirsty insects left us alone. Entering the woods, we noticed ground-hugging purple saxifrage and moss campion carpeting the forest floor as we gained elevation.

We stopped beside a gurgling brook for refreshing drinks of cold clear water.

"Ah, nectar of the gods!" Gil smacked his lips. Near our feet we noticed some very small flowers blooming over a star-shaped arrangement of leaves.

"Those aren't violas." I tried to get a close look.

An incubating yellowlegs came off her nest. Four splotched eggs could be seen in the nest.

Top: *We found a rare insectivorous plant, butterwort. Insects were snared in its sticky leaves.*

Above: *Moose nugget moss grew in velvety clumps about five inches in diameter.*

"Notice that the pale yellowish green leaves look sticky," Gil pointed out. "There are minute insects stuck on the leaves."

We had just found butterwort, a rare, insectivorous plant. It grows in wet places where nitrogen has been leached from the soil. The plant absorbs nitrogen compounds from the insects which become snared in its sticky leaves. But before they are trapped in the gluelike substance the insects have helped pollinate the flower.

Gil and I followed the little brook between Ram Peak and Lamb Butte, climbing to find a breeze to cool us on this warm afternoon. There was a giant plant, big saxifrage, which some large animal, perhaps bear, had been eating A gray-headed chickadee flew downhill with a beakful of insects for its young.

Continuing our ascent we found another unusual plant, two velvety clumps of green moss about five inches in diameter, studded with spore capsules with creamy bell-shaped skirts below; they looked like tiny Chinese parasols supported on wirelike bronze stems, which glistened in the sun. Gil and I found this species of moss growing in only four widely scattered places in the Brooks Range; each time we found it growing on decayed moose pellets. Spores for reproduction are carried by the flies which visit the dung. We called it moose nugget moss. Just below the tree line we stopped beside the stream for supper. We made a small Eskimo stove, placing stones the proper distance apart to support a three-pound shortening can, with a small fire between the rocks to heat water.

Shortly after we entered the woods we found this morel, a mushroom.

A red squirrel climbed a nearby spruce to investigate the strangers who had invaded his territory. A curious chickadee watched while we ate supper of canned tuna, cold potato cakes, hot milk and tea. Dessert was dried fruit. We felt resuscitated. The brook splashed down over clean, mossy rocks. Dryas, lichens and moss cushioned where we lounged, watching the dark rain squalls over the Arrigetch Peaks.

Ramparts of yellowish rocks towered above. Gil spotted a bear den, below the rocks at about 3,200 feet elevation. A hole in the mountainside, with a mass of loose dirt below, showed the den must be of good size. It was far out of our way, so we had no time to inspect the den.

Resuming our climb, we noted another lousewort with pale golden harps beside what resembled ripe wine-colored berries, mountain valerian in bud.

At 2,500 feet in elevation, Gil spotted a bulky nest in a spindly, 10-foot-high spruce. The nest was placed seven feet up in such a precarious position, on such weak branches, that it had tipped and three young birds were lying on the ground dead. Sitting on the tipped nest, a female varied thrush was brooding over one fuzzy survivor. The male joined her in cries of distress when we approached. The scrawny spruce was much too wobbly to climb so I stood on Gil's shoulders to tie the nest securely with an extra boot lace I had in my pack. The parents were wild with excitement, obviously thinking that we were going to murder their last infant. As we moved away the mother hovered over the nestling and settled down to cover it again.

"It was a stupid place to build a nest." I was glad we could help.

We filled our canteens where coltsfoot bloomed abundantly, then ascended the slope through moist grassy hummocks to the Umbilical Ridge, which joins Lamb Butte to the parent peaks, Ram and Ewe. Glacier avens provided spots of bright yellow sunshine and club moss grew just below the crest of the dry ridge at 3,000 feet. We reached the ridge top at 8:00 P.M.

In a few minutes, we made another Eskimo stove and, with a few small willow twigs, made a fire to heat water for a meal of hot muesli, fruit and cocoa. We sat on a wilderness ridge top eating our evening meal, at peace with ourselves and the entire world, while varied thrushes and robins rendered a concert from the spruce far below. It

was calm and beautiful, with the sun fairly high over peaks on the horizon to the northwest. The Arrigetch Peaks dominated the southern skyline; jagged and fearsome granite for climbing, they were lovely gray-streaked spires in the evening light.

Gil and I rested for two hours, gazing all around

We made an Eskimo stove and heated water for a meal of hot muesli, fruit and cocoa.

at the rugged mountain slopes, habitat for moose, bear and wolf, down there, somewhere. When a vagrant breeze came up, we donned long johns and wool undershirts.

About ten o'clock, we continued up Umbilical Ridge, toward the higher rim which connects Ram and Ewe peaks, and followed the undulating ridge up and down and up again to the summit of Ewe, an outpost overlooking the wide valley. Looking down we saw the tortuously winding channels snaking down between precipitous mountain slopes. Numerous lakes glistened on both sides of the main river with its many intertwining channels, a braided stream. We heard a golden plover call a couple of times.

On the summit delicate black oxytrope bloomed in pretty, ground-hugging clumps. The highest block on the summit was beautifully decorated with flat, crustlike cream-and-black lichens. Reindeer moss lay thick between the stones. Apparently this summit had not been desecrated by man.

Since we wished to be on the main ridge by midnight we wended our way back over the ridge. Climbing the last steep pitch, skirting a vertical cliff, we heard "Tee-hee-hee Tee-hee-hee." It sounded like derisive laughter.

"Upland plovers! I'll be darned!" Above us a pair of the elusive large shore birds were sneaking around in the low vegetation. We crouched watching, determined to keep an eye on both birds so we could find their nest.

"This is the sixth time we've encountered these birds at nesting time — all in the Brooks Range." Gil was keeping his binoculars focused on one. "They always have outwitted us when we have looked for their nests. It's their tittering that gets to me. It makes me feel ridiculous."

"Maybe, this time we'll have better luck." I kept my glasses on the other plover, as both birds furtively walked around, feeding and occasionally putting their heads up, keeping watch on us. When we decided to climb up toward them, they flew over the ridge. Darn!

It was time to go up to the summit for pictures of the midnight sun. We could return later to look for plover nests. Climbing up to where the plovers had disappeared, we found both of them in company of a golden plover. All seemed to be defending their territories.

"I'm going up to the summit to locate the sun." Gil moved on. "Why don't you watch the plovers?"

We were in the shade as the sun streaked over the ridge from the north. I used binoculars and watched the upland plovers slink back and forth, until both disappeared into a narrow band of shin-high willows. Then I climbed up to join Gil on the ridge. It was midnight, summer solstice.

The sun was a ball of fire over the mountains due north. Purple peaks studded the horizon. We watched the sun, sitting on that wild lonely ridge to eat our midnight lunch. Upland plovers skittered around in the short shrubs directly below, blending in with the dun-colored tundra. The temperature was near freezing. A slight breeze made us chilly so we donned rain parkas and chaps.

June 21, summer solstice, about midnight. Viv gazed toward the Arrigetch from Ewe Peak.

Arctic lupine grew in profusion in our wilderness.

125

Just after midnight the sun hid itself for half an hour behind a distant spire. An aura of mystery prevailed as we waited for the sun to reappear, then it followed the horizon to the right. Pink puffs of clouds hung over the Arrigetch. Each peak seemed to be an active volcano, puffing out pink smoke. Only the sharp spires seemed to hold the cloud puffs. The sun shone on mountains all around the horizon. Valleys going north-south were bathed in sunlight, while those oriented east-west lay in deep shade. No clouds were visible except for the little pink puffs over the Arrigetch.

When the sun came around the narrow peak we felt its warmth. Gil took photos of the blazing orb slowly circling low on the skyline. It was now shining on Europe and on us at the same time. (When it's midnight in Alaska it is midday in Sweden.)

Returning to our search for the upland plover nest, I went down the slope to the west, traversing toward the suspected nesting area, while Gil stood watch on the ridge. I flushed one of the plovers. It lowered its head, slipping through the vegetation, then flew below. Gil came down to me. We thought we had the nest located. But none was to be found. Frustrated with the upland plovers we turned our attention to the golden plovers. For a half-hour, we lay watching through binoculars. But we had to give up on the goldens, too. We climbed back up to the summit for a snack. Gil's watch said 5:00 A.M.

Heading down the mountain, we again tried to flush an upland plover, but saw none. Traversing toward Umbilical Ridge we heard a golden-crowned sparrow singing his song, "Three Blind Mice." He was in knee-high willows at about 3,300 feet elevation, below the rim, high above spruce.

We stopped on the Umbilical Ridge long enough to dismantle the Eskimo stove we had made, placing the stones exactly where we had found them. The sun shone on Lamb Butte, lower than the summit ridge to the north, as we traipsed over the top of the butte and began a steep descent through short dwarf birch on good wildlife trails. Songs from varied thrush and robins greeted us from scattered spruce above tree line. Fox sparrows and white-crowns joined in.

Soon Gil led down the steep slope through thick head-high dwarf birch which entangled our feet.

*Just after midnight
the sun hid itself
behind a distant spire.
We waited for it to reappear.*

*No clouds were visible
except the little pink puffs
over the Arrigetch.*

Hawk's beard was abundant on the river bars.

"This would be some place to meet a bear!" I could barely see Gil's pack with rifle barrel protruding through the thick brush ahead. I dogged his heels. Fearing we might surprise a grizzly, I began to sing loudly. A few places where water seeped out, our feet slipped out from under us. Reaching the spruce, we passed a robin's nest at the tree line. I looked in to see three fuzzy little heads in a row, snuggled down tightly in the mud-lined nest.

"It's strange," I commented, "that robins in the Arctic use no soft materials to line their nests. It seems to me that the babies would need that insulation in this cold climate."

The brush opened up in the wooded slope and it was a great relief when the trail leveled off. We were tired from our long journey. American red currant was blooming in the woods where we stopped to remove our long johns and dope up with bug repellent.

On the North Creek gravel bars hawk's beard was abundant. They were like plump pincushions of green lace, embroidered with yellow French knots. Along the river we found fresh tracks from a black bear in the sand and mud, heading upstream in the tracks we had made yesterday. As we went along we saw that the bear had followed our tracks. Where had he come from? Had he visited our cabin? Was he the same animal who broke in?

Reaching the cabin at 9:00 A.M., we were relieved to find no evidence that the bear had been there. According to Gil's pedometer we had hiked and climbed 25 miles.

"We were lucky," Gil remarked. "It was the only day and night that it was clear enough to take pictures." We had rain squalls the following night. But we didn't care. We slept most of the day after our climb, then strolled around a bit and went back to bed as the Swainson's thrush serenaded from the island.

•

We had not been able to cross the rivers since breakup. We were eager to get across North Creek to explore the ponds we had seen from Lamb Butte, but flood waters had prevented further investigation.

June 25. We decided to take advantage of beautiful weather and try to take the long-delayed hike, even though North Creek was still rather high.

Gil waded across in hip boots, with his pack and rifle.

"This crossing is tricky," he complained as he negotiated the swift river. I laughed at his concern.

"Don't laugh," he frowned. "If I dunk the cameras it would be disastrous."

Coming back for me he had trouble keeping his footing. The fast water nearly came over his boot tops. Standing beside me, he watched the rushing river dubiously. "I just about went down."

"What should we do?"

"Oh, well, let's give it a whirl." I was wearing my loaded pack when Gil picked me up for a piggyback ride over the turbulent stream. It was shallow at first, then got deeper. Halfway across, Gil stood with legs braced far apart. My steed was unsteady, taking small steps, moving slowly over the round boulders beneath. When we made it safely through the deep water we heaved sighs of relief.

"I hope the water doesn't come up today or we're in trouble. I don't mind getting wet, but I'm concerned about the cameras."

Gil changed to shoepacks and we headed upstream on the opposite bank of North Creek. It was a refreshing change for us.

Walking on a good wildlife trail near the river, we saw the first prickly wild rose of the season, blooming on the riverbank, as large as a teacup. Shafts of sourdock were spectacular. Crowberry was loaded with green berries. A robin agressively defended his territory from a young gray jay and a Swainson's thrush slipped off her nest, a grass cup, decorated with reindeer moss; it was about knee-high in a small spruce. We counted four wobbly fuzzy heads inside.

I filled a canteen before we left the river and then we headed north on slightly higher ground. Gil was photographing some lavender fleabane when he noticed several gold-colored lady's slippers. Delighted with this most beautiful orchid, only four inches tall, the photographer lay on the ground, focusing the camera on three golden slippers all in a row. Later, he discovered some of the lady's slippers in white.

Tramping through the woods, Gil was leading when he surprised a bull moose, a beautiful animal, fat and sleek, with a wide spread of antlers in velvet. The moose seemed curious and stood for some pictures before he swung around to stride

regally over a hill. We followed in the footsteps of the big bull, proceeding to a prominent bluff overlooking a small lake, arriving just in time to see moose antlers disappear underwater in the middle of the pond. All that was visible was a series of concentric rings to mark where the beast had submerged. Quietly we waited on the bluff. Then the moose head rose like a submarine in the middle of the little lake. For nearly an hour we sat watching the moose submerge, raise his head and go down again. He shared the pond with a family of ducks. A female widgeon swam on the lake, followed by a string of eight downy babies. Not bothered or unafraid of the moose, they swam close to him.

Wandering toward the big river to another small

Three gold lady's slippers, only four inches tall, glistened in the sunlight.

Gil surprised a bull moose while tramping through the woods.

pond we were escorted all the way by a pair of yellowlegs with their tiresome loud crying. Those cries without letup got on my nerves so much that I felt like throwing something at them.

Arctic terns protested our presence with soft rasping calls. We saw their downy chicks bobbing out on the water, like brown and gold corks. A pair of horned grebes claimed the pond. One of the grebes swam to the far side and climbed up on a fallen spruce lying in the water. Her nest was cradled among the branches. She began uncovering her eggs before settling to incubate. Waxwings and pine grosbeaks flew in and out of the spruce on shore.

"Look at that!" Gil pointed to the tern family. "The parents hover over the chicks and feed them while they're swimming!"

Before leaving the pond, I climbed a tree to count five eggs in the grebe's nest, 30 feet from shore. The persistent yellowlegs escorted us from the pond. We were glad to leave.

Sweet vetch bloomed profusely on the shore of the big river, perfuming the air as we followed the stream back toward home.

"I want to carry you back across the river while I still have the strength." Gil was weak with hunger.

The water in North Creek was down slightly when Gil piggybacked me across. We had made a large triangle, hiking 10 miles in eight hours.

•

June 28.   Three mew gulls and four terns were flycatching low over the big river, harvesting a batch of large insects, making "pablum" for their young.

Carrying a rifle and Gil's camera gear, we hiked out to the Enchanted Forest for a final check on the progress of Three-toes's young. When we were about 200 feet away we heard the rapid, rattling call of the youngsters, who seemed about ready to fly. Each wore a gold beanie. We had no way to count the noisy nestlings.

"Everyone in the family wears a yellow beanie except the mother," I pointed out to Gil. "The young females will lose the cap when they reach breeding age."

Gil and I went on to visit the goshawks and were met by a sudden angry swoosh over our heads. "Kyak-kyawk," cried the agitated parent as it perched above the nest to glare down at us. A downy chick lay sprawled on the edge of the stick platform, watching us intently.

"That chick must be about two weeks old," Gil surmised.

The baby hawk did not move a muscle all the time we were there, only following us with its eyes, peering through the branches which supported it. We could not see if there were any more young in their lookout post. The adult remained perched above, keeping watch. I moved around to the opposite side of their tree. Again the enraged parent dropped down with wild cries in a frenzied swoop before it perched once more to watch. Not wishing to disturb them further, we left the Enchanted Forest.

Back at the cabin, we decided to check on the

kestrels' progress. Using the cache ladder, we climbed up, opened the lid and found five downy young a few days old. The soft cuddly babies were lying flat and still when each of us reached in to count the newborn.

"Now the kestrel has six mouths to feed," Gil noted. "That'll keep him busy."

June 29. *After supper we strolled out to the bear trail and heard a loud, resonant, gnawing sound. It was a porcupine chewing on the moose antler. Mosquitoes were bothersome. The porky shook his head, waved his forepaws over his face and scratched his tummy where the fur is thin. Finally he decided the photographer was too close. He waddled away.*

We heard a loud, resonant sound. A porcupine was gnawing on an antler.

131

13 · Farewell, Wilderness

Gil and I planned to start floating out to civilization on July 4. In spite of the fact that we had been hungry since January, we kept to our schedule. We desperately wanted to finish our work of recording nests and other data before leaving. By stringent rationing we stretched our meager food supply enough to make it. We lost weight steadily and were mere shadows of our former selves when, on July 2, it was time to start packing up to leave the Arctic.

Gil climbed the ladder to the cache to start dismantling it. The decking was lashed to crossbeams, which were lashed to trees with light nylon line. As Gil loosened the poles he tossed them down and I stacked them in a neat pile by the cabin. Tins were removed from the supporting "legs" to reveal that the beautiful old spruce which we had used were undamaged, except for a few small branches which we had trimmed off.

Our permit had allowed us to build the cabin on land administered by the Bureau of Land Management and now our time in the wilderness had run out. We had signed an agreement to dismantle the cabin before we left. We knew we would be subject to a sizable fine if we didn't abide by the agreement.

"But," said Carl Johnson, the district manager, as we parted last fall, "I'm hoping you won't have to dismantle it. Don't start taking it apart until you hear from me."

But we weren't getting any mail. We wondered what to do. We loved the cabin. We had spent the happiest year of our lives here. It would tear our hearts out to take it apart, but we would if we had to.

"If we take it apart and find out later that we could have left it standing, it would be a blow." My heart was already aching at the thought of leaving the wilderness.

"Let's leave it standing," Gil suggested. "When we reach civilization if we learn that we should have dismantled it we can hire someone to come back to do it." That settled the matter. We felt better as we prayed the cabin would be saved.

Viv gazed at the cabin trying to fix in her mind every precious detail. Then Gil fastened the door shut.

July 3 was a busy day packing, sorting, getting ready to leave. We burned some of the trash. Tins would be left in the cabin for rodent-proof storage for future visitors to use.

It took Gil a half-hour to blow up the rubber raft by mouth. The golden-yellow color was a spot of sunshine in the yard. "We'll call it *Golden Goose*," Gil announced. With a flow pen, he printed the name on each side of the bow.

A downpour of rain came in the afternoon and it looked too bleak to start out the next day, so we splurged and each had a cup of coffee to perk up our spirits, while listening to the patter of little feet on our rooftop. It was one of Snickaree's kids, a tiny red squirrel. A pine grosbeak sang some sweet notes and the mews cried in alarm and pursued some source of danger to their children.

We continued packing. What were we to do with our important papers, books and records? I put them in sealed plastic bags in my canvas pack with our diaries as my responsibility. Gil had the responsibility for cameras and film. A year's work would be wasted if these records were lost in the river.

July 3. *I have a strange, sickish feeling in the pit of my stomach at the thought of leaving our little wilderness home.*

July 4. *I awoke. Then I realized it is the day of departure. And cry. When I looked at the rough ceiling poles of this dear cabin, my heart is heavy at the thought of leaving. Sad to leave all the beauty we have found here. Gil is cooking breakfast on the stove. Flowers still on the table in a clear vase made from a plastic bottle. A lovely wild rose, some frothy white valerian and Jacob's ladder. But we must go, and mush is cooked, so I must get up from this dear bed, which is rated the "hardness of 3" as geologists measure hardness of rock.*

A bear had been close to the cabin while we slept. He left a calling card in the trail, a large dark scat. Luckily, Gil had put the rubber raft on the roof overnight for safekeeping.

Gil and I worked steadily all morning finishing our packing. No more time to be sorry about leaving — too busy! Gil brought the rubber boat down off the roof. We carried out the bulk of our packages to plan how they could be fitted into the seven-man raft.

"You'll never get all those things in." I had never floated a river before and was apprehensive.

"We can't leave any of this behind." Gil was firm. "That five-gallon tin holds our exposed film. Our packs, sleeping bag, tents, clothing — what's left of it — camp stove, cooking pots, rifle and cameras, all have to go with us."

"We don't have room for the snowshoes." I started to carry them back into the cabin.

"Those are borrowed, remember?" He carried them back out.

"There'll be no room for us."

"We can stack things high."

I could see it now. The rubber bottom of the boat would bulge down dangerously.

"We'll scrape bottom on rocks and tear a hole in

it." I was worried. "Those first rapids look pretty wild. Remember, we have 200 miles to go."

Compromising, we agreed to take everything down to the big river, a half-mile away, and load up before final decisions were made.

We tidied up the cabin. Gil hand-lettered a sign on the inside of the cabin door. PROPERTY OF U.S. GOVERNMENT, DEPT. OF B.L.M. We boarded up the windows with heavy shutters we had made.

I went back inside the cabin and my eyes swept over the beautiful, rugged interior as I tried to fix in my mind every precious detail. Then Gil fastened the door shut with a heavy wire.

We began relaying all our gear, baggage and boat out toward the river. When everything was halfway we returned to the cabin to say good-by to it.

My eyes filled with tears as I looked at it for the final time. It was harder to leave this crude, moss-chinked cabin than our fine home in Oregon. How quaint the little hut looked in its wilderness setting!

A pine grosbeak sang his full spring song as we walked out through the pink fragrance of sweet vetch on the bars. We turned back for another long look. The male kestrel was calling chow time for his family. A dark gray young jay flew across to the island, mimicking the call notes of the Swainson's thrush. The thrushes chased the impudent young jay from the willows, where their nestlings were hidden.

Gil carried the empty rubber raft over his head as I helped with other bundles. At last all our baggage was on the gravel bar, at the jumping-off place. The two mew gulls, who nested upstream, came to investigate the activity. They settled down a hundred feet away to watch us.

"It's nice to have them see us off." I was pleased at their attention.

Gil launched the *Golden Goose* into the river and anchored it to some driftwood so we could begin loading. We spent two hours loading, unloading, rearranging, and trying to fit things in. All our bundles were lashed down and each item

Viv worked steadily to finish the packing. Supplies were stacked in the Golden Goose.

135

had a separate line fastened to a main line, which encircled the boat through grommets. In case the raft punctured or capsized, all our bundles would be strung together and could be retrieved, even if wet. Many items were sealed in plastic bags to prevent water damage. Finally, at Gil's insistence, we got everything aboard the *Golden Goose*. It was stacked high. By late afternoon we were ready to shove off.

"We're overloaded. We'll scrape bottom," I complained, as I started to get aboard. "There's no room for my feet inside."

"That's O.K.," Gil replied, ever the optimist. "You can dangle your feet over the side."

"There's not enough room for you either. Where will you sit?"

"We'll manage. Don't worry!" Gil climbed aboard, finally found room for his legs, and started untying the anchor lines.

"When floating downstream," Gil informed me, "you head the boat upstream."

"Now, wait a minute," I sputtered. "That makes no sense."

"That's what I was told." He had never floated before either. Because he would man the oars he won the argument.

For a year I had worried about this 200-mile float trip. We had discussed earlier whether we should carry all our gear and the raft a half-mile farther downriver to avoid some wild water with rocks jutting out in the middle of the river. After studying the rapids we saw the smoothest water was on the left side, going down. We decided that was the safest water to negotiate with our heavy load.

We shoved off. Gil started rowing frantically to try to control our craft, to keep us from drifting down the slough. We made it into the main current, then deep water. Then a whirlpool took the raft out of control, we bumped the opposite shore, and hung up.

"I've already lost confidence in your oarsmanship." I wondered if we'd ever make it.

We shoved off again across from where we started. I held on tightly as we were swept toward the rapids. Gil pulled hard on the oars.

"I thought we were going on the left side of the rocks."

"I'm trying." But we drifted to the right. Gil was rowing desperately. "I can't control it."

We spun out of control and were dragged inexorably into the current leading toward the right side of the jutting rocks. I hung on for dear life.

"We're committed now." Gil looked grim. "Here we go!" He fought to keep the raft away from some snags which had fallen into the stream. "Hang on!" We rushed into the boiling turmoil. The *Golden Goose* lurched, buckled in the middle and shuddered as we galloped through the rapids."

Later, in calmer water, we grinned at each other. It had been fun. "Whew!"

It took 11 days to travel the 200 river miles to the nearest Eskimo-Indian village.

Farewell, Wilderness

14·Return to Oregon

Floating down that arctic river was the perfect way to exit from the wilderness year, which now seems like a beautiful dream. Holding still in the raft, Gil and I observed a fox, with a mouthful of rodents which he was carrying home to his family. Wolves lounged on the islands, watching us glide silently past. Moose stared, too, without fear.

When we landed at Allakaket, Natives politely asked us where we came from — as if they didn't know. It was their gentle way to open the conversation. Actually they had been worried about our survival up there.

The Wien plane to Fairbanks was due the next morning. When it arrived, we saw that it was built like a big square box, with small wings. I asked the pilot, who wore his uniform cap at a jaunty angle, if that thing could fly.

"Lady, if you show me the way to go, we'll fly to Bettles."

Later, at Bettles, there was a layover while the pilot took a side trip up to Anaktuvuk Pass and back. I asked him when we should be ready to board the plane for Fairbanks. While looking back at me, he walked toward the log lodge, and answered, "In two hours, be ready." Then not looking where he was going, he dropped into an excavation for a new cellar for the lodge. He climbed out, looking sheepish. "Don't mind me. I always do that when I look at a woman," he said.

Gil and I had lunch with Jeannie and Daryl Morris, the bush pilot who had flown us into the wilderness. They had a letter for us, mailed the year before, from Carl Johnson, of the BLM. The letter said, "Please don't dismantle your cabin." There was only time for quick good-bys to our friends before we ran to catch the plane.

On the aircraft, getting ready to take off, the pilot counted the passengers. "Too many aboard. We'll be overloaded." Taking a roll call, he found the extra man, who had been told he would have to wait for the next flight. There was an argument. Finally the pilot let him stay on. Approaching the Fairbanks airport, the "beeper" started signaling stall speed while we were still 3,000 feet up. Gil and I were really worried when the beeper kept starting up, while we descended at a steep angle — too fast. We slammed down hard on the runway and the flying boxcar went into a terrible shimmy. Up front we saw the pilot and co-pilot bouncing around wildly, while the pilot gripped the wheel. The whole plane was vibrating violently, like a dog shaking himself after a swim. The pilot got the plane settled down just before the end of the runway, where we made a fast turnaround, nearly tipping over.

"They say it's a good landing if you can walk away from the plane," one of the passengers remarked dryly. We learned later that the hard landing had broken something in the front landing gear.

•

Exactly 36 hours after our wilderness experience had ended when we reached that first Eskimo-Indian village, Gil and I found ourselves at Portland International Airport, suffering from severe cultural shock. A friend drove us to Gil's parents' home in the city, via a six-lane freeway.

Had the whole world gone crazy? Where was everyone driving in such a hurry?

Gil's parents were not at home. Confused and exhausted, we went to bed but could not sleep. The noise from a nearby freeway kept us awake. And the bed was too soft, so we made our bed on the floor, but still we could not sleep. We took some tranquilizers and finally dozed off.

"Honey, a plane is buzzing the cabin!" I sat up when the roar of a motor woke us. (I thought we were still in the wilderness.) It turned out to be a big truck shifting gears on the freeway.

We couldn't wait to get away from the hubbub of the big city to our own little wilderness at Wildhaven in the Deschutes Forest of central Oregon. It was good to be back, camped in the forest, even though we had no house, but we had a tent with space, clean air, peace and quiet, and birds and deer for company.

There in the pine forest, as Gil and I reflected on the incredibly beautiful arctic year, we prayed that there would always be wilderness where future generations could find solace and inspiration.

Epilogue

In July 1982 our wilderness home, Wildhaven, was formally donated and dedicated to The Nature Conservancy. We have retained a "life estate," which means we can live here the rest of our lives as caretakers. That is how we regard ourselves anyhow. We live in Nature House, a hand-built stone house, with no electricity, no phone or plumbing. We heat and cook with wood.

Here at Wildhaven, our lives revolve around environmental education, as we present narrated slide programs, lead nature trips and teach at outdoor schools. In addition, we host Natural History weekends for groups from schools, Audubon societies, the Oregon Museum of Science and Industry and other organizations.

Gil and I like to share Wildhaven. For the least disturbance to the wildlife, no motor vehicles are allowed on the last quarter-mile up to Nature House except for emergencies. So visitors must walk up to enjoy. Nicholas Drapela penned his appreciation in our guest book as follows: *Words cannot express my feelings here at Wildhaven. It is a natural beauty mark of the earth, a tranquil shelter from hostility, an everlasting landmark of peace and grace.*

Viv + Gil Staender

Vivian and Gil Staender
Sisters, Oregon
November 1982

Nature House at Wildhaven, our home in Oregon.

140

Birds Identified

Within a five-mile radius of our cabin from July 1969 to July 1970

Single asterisk * signifies nest record for species
Double asterisk ** means evidence of nesting

Loon Family (Gaviidae)
Red-throated Loon *(Gavia stellata)*
Arctic Loon *(Gavia arctica)*

Grebe Family (Podicipididae)
Horned Grebe* *(Podiceps auritus)*

Swan, Goose and Duck Family (Anatidae)
Greater White-fronted Goose *(Anser albifrons)*
Snow Goose *(Chen caerulescens)*
Canada Goose *(Branta canadensis)*
Green-winged Teal *(Anas crecca)*
Mallard *(Anas platyrhynchos)*
Northern Pintail *(Anas acuta)*
Northern Shoveler *(Anas clypeata)*
American Wigeon* *(Anas americana)*
Harlequin Duck *(Histrionicus histrionicus)*
Bufflehead *(Bucephala albeola)*
Red-breasted Merganser *(Mergus serrator)*

Hawk, Eagle and Harrier Family (Accipitridae)
Bald Eagle *(Haliaeetus leucocephalus)*
Northern Harrier *(Circus cyaneus)*
Sharp-shinned Hawk *(Accipiter striatus)*
Northern Goshawk* *(Accipiter gentilis)*
Golden Eagle* *(Aquila chrysaetos)*

Falcon Family (Falconidae)
American Kestrel* *(Falco sparverius)*
Merlin *(Falco columbarius)*
Peregrine Falcon *(Falco peregrinus)*
Gyrfalcon *(Falco rusticolus)*

Grouse, Pheasant and Quail Family (Phasianidae)
Spruce Grouse** *(Dendragapus canadensis)*
Willow Ptarmigan *(Lagopus lagopus)*
Rock Ptarmigan *(Lagopus mutus)*

Plover Family (Charadriidae)
Black-bellied Plover *(Pluvialis squatarola)*
Lesser Golden Plover** *(Pluvialis dominica)*
Semipalmated Plover *(Charadrius semipalmatus)*
Eurasian Dotterel *(Charadrius morinellus)*

Sandpiper Family (Scolopacidae)
Lesser Yellowlegs* *(Tringa flavipes)*
Solitary Sandpiper** *(Tringa solitaria)*
Wandering Tattler *(Heteroscelus incanus)*
Spotted Sandpiper *(Actitis macularia)*
Upland Sandpiper** *(Bartramia longicauda)*
Semipalmated Sandpiper *(Calidris pusilla)*
Baird's Sandpiper *(Calidris bairdii)*
Pectoral Sandpiper *(Calidris melanotos)*
Dunlin *(Calidris alpina)*
Long-billed Dowitcher *(Limnodromus scolopaceus)*
Common Snipe** *(Gallinago gallinago)*
Red-necked Phalarope** *(Phalaropus lobatus)*
Red Phalarope** *(Phalaropus fulicarius)*

Gull, Tern and Jaeger Family (Laridae)
Long-tailed Jaeger *(Stercorarius longicaudus)*
Bonaparte's Gull *(Larus philadelphia)*
Mew Gull* *(Larus canus)*
Herring Gull *(Larus argentatus)*
Glaucous Gull *(Larus hyperboreus)*
Arctic Tern* *(Sterna paradisaea)*

Owl Family (Strigidae)
Great Horned Owl** *(Bubo virginianus)*
Northern Hawk-Owl** *(Surnia ulula)*
Short-eared Owl *(Asio flammeus)*
Boreal Owl* *(Aegolius funereus)*

Kingfisher Family (Alcedinidae)
Belted Kingfisher *(Ceryle alcyon)*

Woodpecker Family (Picidae)
Three-toed Woodpecker* *(Picoides tridactytus)*
Northern Flicker* *(Colaptes auratus)*

Tyrant Flycatcher Family (Tyrannidae)
Olive-sided Flycatcher *(Contopus borealis)*
Western Wood-Pewee *(Contopus sordidulus)*

Lark Family (Alaudidae)
Horned Lark *(Eremophila alpestris)*

Swallow Family (Hirundinidae)
Tree Swallow *(Tachycineta bicolor)*
Violet-green Swallow* *(Tachycineta thalassina)*
Bank Swallow* *(Riparia riparia)*
Cliff Swallow *(Hirundo pyrrhonota)*

Jay, Magpie and Crow Family (Corvidae)
Gray Jay *(Perisoreus canadensis)*
Common Raven *(Corvus corax)*

Titmouse Family (Paridae)
Siberian Tit *(Parus cinctus)*
Boreal Chickadee* *(Parus hudsonicus)*

Dipper Family (Cinclidae)
American Dipper *(Cinclus mexicanus)*

Old World Flycatcher, Warbler and Thrush Family (Muscicapidae)
Arctic Warbler *(Phylloscopus borealis)*
Ruby-crowned Kinglet *(Regulus calendula)*
Northern Wheatear* *(Oenanthe oenanthe)*
Gray-cheeked Thrush* *(Catharus minimus)*
Swainson's Thrush* *(Catharus ustulatus)*
American Robin* *(Turdus migratorius)*
Varied Thrush* *(Ixoreus naevius)*

Wagtail and Pipit Family (Motacillidae)
  Water Pipit *(Anthus spinoletta)*
Waxwing Family (Bombycillidae)
  Bohemian Waxwing *(Bombycilla garrulus)*
Shrike Family (Laniidae)
  Northern Shrike* *(Lanius excubitor)*
Wood Warbler, Tanager, Grosbeak, Bunting and
Blackbird Family (Emberizidae)
  Orange-crowned Warbler *(Vermivora celata)*
  Yellow Warbler *(Dendroica petechia)*
  Yellow-rumped Warbler* *(Dendroica coronata)*
  Northern Waterthrush** *(Seiurus
    noveboracensis)*
  Wilson's Warbler *(Wilsonia pusilla)*
  American Tree Sparrow *(Spizella arborea)*
  Savannah Sparrow* *(Passerculus sandwichensis)*
  Fox Sparrow *(Passerella iliaca)*
  Golden-crowned Sparrow** *(Zonotrichia
    atricapilla)*
  White-crowned Sparrow* *(Zonotrichia
    leucophrys)*
  Dark-eyed Junco* *(Junco hyemalis)*
  Snow Bunting *(Plectrophenax nivalis)*
  Rusty Blackbird *(Euphagus carolinus)*
Northern Finch Family (Fringillidae)
  Pine Grosbeak *(Pinicola enucleator)*
  White-winged Crossbill *(Loxia leucoptera)*
  Common Redpoll *(Carduelis flammea)*
  Hoary Redpoll *(Carduelis hornemanni)*

Wandering tattler
(Heteroscelus incanus), *a*
member of the sandpiper
family (Scolopacidae).

Plants Identified

Within a five-mile radius of our cabin from July 1969 to July 1970

Cladonia Family (Cladoniaceae)
  Reindeer moss *(Cladonia rangiferina)*

Splachnum Family (Splachnaceae)
  Moose nugget moss *(Splachnum luteum)*

Club Moss Family (Lycopodiaceae)
  Club moss *(Lycopodium selago)*

Horsetail Family (Equisetaceae)
  Common horsetail *(Equisetum arvense)*

Cypress Family (Cupressaceae)
  Common juniper *(Juniperus communis)*

Pine Family (Pinaceae)
  White spruce *(Picea glauca)*

Sedge Family (Cyperaceae)
  Cotton grass *(Eriophorum vaginatum)*

Lily Family (Liliaceae)
  Death camas *(Zygadenus elegans)*

Orchid Family (Orchidaceae)
  Gold lady's slipper *(Cypripedium calceolua)*
  White lady's slipper *(Cypripedium montanum)*
  One-leaved rein-orchid *(Habenaria obtusata)*
  Slender spire orchid *(Habenaria unalascensis)*

Willow Family (Salicaceae)
  Quaking aspen *(Populus tremuloides)*
  Balsam poplar *(Populus tacamahaca)*
  Feltleaf willow *(Salix alaxensis)*
  Snow willow *(Salix reticulata)*

Birch Family (Betulaceae)
  Shrubby alder *(Alnus crispa)*
  Dwarf birch *(Betula nana)*
  Paper birch *(Betula papyrifera)*

Buckwheat Family (Polygonaceae)
  Mountain sorrel *(Oxyria digyna)*
  Pink plume *(Polygonum bistorta plumosum)*
  Arctic sourdock *(Rumex arcticus)*

Pink Family (Caryophyllaceae)
  Moss campion *(Silene acaulis)*

Buttercup Family (Ranunculaceae)
  Monkshood *(Aconitum delphinifolium)*
  Larkspur *(Delphinium brachycentrum)*
  Purple pasque flower *(Pulsatilla ludoviciana)*
  Buttercup *(Ranunculus* species*)*

Mustard Family (Cruciferae)
  Parry's wallflower *(Parraya nudicaulis)*

Saxifrage Family (Saxifragaceae)
  Purple saxifrage *(Saxifraga oppositifolia)*
  Big saxifrage *(Therofon richardsonii)*

Rose Family (Rosaceae)
  Yellow dryad *(Dryas drummondi)*
  White dryad *(Dryas octopetela)*
  Glacier avens *(Geum glaciale)*
  Tundra rose *(Potentilla fruticosa)*
  Prickly wild rose *(Rosa acicularis)*
  Nagoonberry *(Rubus acaulis)*

Pea Family (Leguminosae)
  Sweet vetch *(Hedysarum boreale mackenzii)*
  Eskimo potato *(Hedysarum alpinum)*
  Arctic lupine *(Lupinus arcticus)*
  Locoweed *(Oxytropis deflexa foliolosa)*
  Black oxytrope *(Oxytropis nigrescens)*
  Vetch *(Vicia* species*)*

Crowberry Family (Empetraceae)
  Crowberry *(Empetrum nigrum)*

Evening Primrose Family (Onagraceae)
  Arctic Fireweed *(Epilobium latifolium)*

Mare's-Tail Family (Hippuridaceae)
  Mare's-tail *(Hippuris vulgaris)*

Heath Family (Ericaceae)
  Pink pyrola *(Pyrola grandiflora)*
  One-sided wintergreen *(Pyrola secunda)*
  Bog rosemary *(Andromeda polifolia)*
  Red bearberry *(Arctostaphylos rubra)*
  Mountain heather *(Cassiope tetragona)*
  Labrador tea *(Ledum decumbens)*
  Lapland rosebay *(Rhododendron lapponicum)*
  Blueberry *(Vaccinium uliginosum)*
  Cranberry *(Vaccinium vitis-idaea)*

Primrose Family (Primulaceae)
  Shooting star *(Dodecatheon frigidum)*

Gentian Family (Gentianaceae)
  Delicate gentian *(Gentiana propinqua)*

Phlox Family (Polemoniaceae)
  Jacob's ladder *(Polemonium acutiflorum)*
  Northern Jacob's ladder *(Polemonium boreale)*

Figwort Family (Scrophulariaceae)
  Pallid paintbrush *(Castilleja pallida)*
  Dwarf lousewort *(Pedicularis capitata)*
  Wooly lousewort *(Pedicularis lanata)*
  Elephant's trunk lousewort *(Pedicularis parviflora)*
  Bladderwort *(Lentibulariaceae)*
  Butterwort *(Pinguicula vulgaris)*

Honeysuckle Family (Caprifoliaceae)
  Twinflower *(Linnaea borealis)*

Sunflower Family (Compositae)
  Alpine arnica *(Arnica alpina)*
  Wormwood *(Artemisia frigida)*
  Siberian aster *(Aster sibiricus)*
  Hawk's beard *(Crepis nana)*
  Fleabane *(Erigeron humilis)*
  Arctic coltsfoot *(Petasites frigidus)*
  Purple hawkweed *(Saussera densa)*

Snow willow pussies (Salix reticulata), *a member of the willow family (Salicaceae)*

Our Food List

We seriously underestimated our food requirements. In parentheses is what we should have taken where experience proved our supplies too short.

50 lbs.	Flour (100 lbs.)	8 lbs.	Instant coffee (15 lbs.)	2 lbs.	Miscellaneous mixes, 23 packages
35 lbs.	Pancake mix	3 lbs.	Tea	5 oz.	Syrup mix, 5 packages
25 lbs.	Biscuit mix (50 lbs.)	3 lbs.	Sweetened cocoa mix	6 oz.	Dried yeast, 24 packages
10 lbs.	Wheat hearts (20 lbs.)	12 lbs.	Canadian bacon, 6 cans	1 lb.	Assorted seasonings
4 lbs.	Rolled wheat	27 lbs.	Bacon, 27 cans	6 lbs.	Salt (10 lbs.)
20 lbs.	Oatmeal (30 lbs.)	8 lbs.	Beef, 4 cans	1 lb.	Soda
3 lbs.	Muesli cereal	2 lbs.	Salmon, 2 cans	½ lb.	Baking powder (1 lb.)
2 lbs.	Bulgur	4½ lbs.	Tuna, 11 cans	15 lbs.	Candy (40 lbs.)
34 lbs.	Dried beans (50 lbs.)	1½ lbs.	Sardines, 8 cans	1½ lbs.	Mixed nuts (5 lbs.)
3 lbs.	Dried peas (10 lbs.)	½ lb.	Crab, 1 can	12 lbs.	Raisins (20 lbs.)
1 lb.	Lentils	¾ lb.	Shrimp, 2 cans	3 lbs.	Figs (5 lbs.)
20 lbs.	Macaroni products (40 lbs.)	6 lbs.	Salami sausage (10 lbs.)	5 lbs.	Pitted prunes (25 lbs.)
50 lbs.	Rice	4 lbs.	Peanut butter (8 lbs.)	4 lbs.	Pitted dates (10 lbs.)
3 lbs.	Potato balls	6 lbs.	Cheddar cheese (12 lbs.)	2 lbs.	Date-nut mix
12 lbs.	Instant potatoes (18 lbs.)	10 lbs.	Powdered eggs (20 lbs.)	6 lbs.	Dried peaches, 2 No. 10 tins
9 lbs.	Assorted crackers	7½ lbs.	Jam, jelly	5 lbs.	Dried apples, 2 No. 10 tins
100 lbs.	Sugar	5 lbs.	Honey (10 lbs.)	8 lbs.	No-bake pie mixes, 15 packages (10 lbs.)
100 lbs.	Powdered milk	6 lbs.	Brown sugar (10 lbs.)		
60 lbs.	Margarine	1 lb.	Powdered sugar (10 lbs.)	8 lbs.	Pudding mixes (20 lbs.)
9 lbs.	Vegetable shortening (12 lbs.)	¼ lb.	Potato pancake mix	4 lbs.	Whipped topping mix (10 lbs.)
1 lb.	Dried onion (3 lbs.)	3½ lbs.	Miscellaneous soup mixes, 24 packages (50 packages)	5 lbs.	Flavored gelatin
2 lbs.	Cocoa			2 lbs.	Graham cracker crust mix
2 lbs.	Coffee creamer	1 lb.	Sour cream sauce mix, 10 packages		

Our Supply List

Kitchen Equipment

2 lbs.	Kerosene stove
240 lbs.	Kerosene (30 gallons)
½ lb.	Aluminum wind screen for stove
3 lbs.	Cook pots and lids
2 lbs.	Plastic bowls
1 lb.	Dishes (2 plastic plates, 2 metal plates, 2 cups)
1 lb.	Knives, forks, spoons
1 lb.	Skillet
1 lb.	Pancake turner, beater, sifter, big spoon
1 lb.	Matches
—	Plastic cleaning pads
15 lbs.	52 bars of hand soap
18 lbs.	Toilet tissue (48 rolls)
4 lbs.	Aladdin kerosene lamp chimneys
1 lb.	Small kerosene lantern

Miscellaneous

20 lbs.	Cameras and attachments: 35mm Pentax, Model H1A; 35mm Praktika; 35mm Lens—400mm, 55mm; 2¼x2¼mm Hasselblad F1000 with two interchangeable backs for black-and-white and color; Hasselblad lens—200mm, 80mm.
15 lbs.	Film (about 140 rolls)
20 lbs.	Firearms and ammunition

5 lbs.	Fishing gear (3 poles)
2 lbs.	First-aid supplies
2 lbs.	Toilet articles
2½ lbs.	Binoculars (2 pairs)
20 lbs.	Books, diaries, paper, etc.
22 lbs.	Car-camping tent, with pegs
7 lbs.	Tarpaulins, poncho
15 lbs.	Air mattresses (1 double heavyweight and 2 single lightweight) with case
5 lbs.	Double sleeping bag
5 lbs.	2 wool blankets
4 lbs.	Mosquito netting, headnets
5 lbs.	Assorted plastic sheeting
2 lbs.	2 burlap bags, for floor mats
8 lbs.	Parachute (Fabric lined the cabin wall behind the bed, and kept the blankets from sticking to moss chinking.)
5 lbs.	Ropes, nylon lines, carabiners
36 lbs.	Drum for heater, stovepipe
8 lbs.	Stovepipe oven
5 lbs.	Window glass (8 panes)
6 lbs.	4½-foot crosscut saw, 2 hand saws
5 lbs.	Double-bitted ax, with extra handle
5 lbs.	Auger (drill) with ratchet
3 lbs.	Crescent wrench, pliers, screwdriver
1½ lbs.	Hatchet (hammer)

12 lbs.	2 pairs snowshoes
9 lbs.	3 backpacks (2 lightweight frame packs, 1 Bergen pack)
5 lbs.	Drill and bit
30 lbs.	Rubber life raft
8½ lbs.	2 pairs heavy hiking boots
7 lbs.	2 pairs shoepacks
6 lbs.	1 pair rubber hip boots
10 lbs.	5 pairs mukluks
2 lbs.	2 pairs light slippers
10 lbs.	20 pairs wool socks; felt boot liners
6½ lbs.	4 pairs wool long johns
5 lbs.	3 pairs heavy jeans
6 lbs.	3 pairs wool trousers
2½ lbs.	1 pair heavy ski pants
2 lbs.	1 pair heavy melton pants
3 lbs.	2 pairs army field pants
1 lb.	1 field pants liner
3 lbs.	Extra light pants and jeans
5 lbs.	2 down jackets
3 lbs.	2 cagoules (big nylon rain parkas) and chaps
3 lbs.	2 big cotton winter parkas
1½ lbs.	2 light parkas
2 lbs.	Gil's field jacket
4 lbs.	2 wool shirts
3 lbs.	2 wool sweaters
2½ lbs.	2 cotton flannel shirts

Index

Alaska Geographic® Back Issues

The North Slope, Vol. 1, No. 1. The charter issue of *ALASKA GEOGRAPHIC®* Out of print.

One Man's Wilderness, Vol. 1, No. 2. The story of a dream shared by many, fulfilled by a few; a man goes into the Bush, builds a cabin and shares his incredible wilderness experience. Color photos. 116 pages, $9.95.

Admiralty . . . Island in Contention, Vol. 1, No. 3. An intimate and multifaceted view of Admiralty; it's geological and historical past, its present-day geography, wildlife and sparse human population. Color photos. 78 pages, $5.

Fisheries of the North Pacific: History, Species, Gear & Processes, Vol. 1, No. 4. Out of print. (Book edition available)

The Alaska-Yukon Wild Flowers Guide, Vol. 2, No. 1. Out of print. (Book edition available)

Glacier Bay: Old Ice, New Land, Vol. 3, No. 1. The expansive wilderness of southeastern Alaska's Glacier Bay National Monument (recently proclaimed a national park and preserve) unfolds in crisp text and color photographs. Records the flora and fauna of the area, its natural history, with hike and cruise information, plus a large-scale color map. 132 pages, $11.95.

Richard Harrington's Antarctic, Vol. 3, No. 3. The Canadian photojournalist guides readers through remote and little understood regions of the Antarctic and subantarctic. More than 200 color photos and a large fold-out map. 104 pages, $8.95.

Southeast: Alaska's Panhandle, Vol. 5, No. 2. Explores southeastern Alaska's maze of fjords and islands, mossy forests and glacier-draped mountains — from Dixon Entrance to Icy Bay, including all of the state's fabled Inside Passage. Along the way are profiles of every town, together with a look at the region's history, economy, people, attractions and future. Includes large fold-out map and seven area maps. 192 pages, $12.95.

Alaska Whales and Whaling, Vol. 5, No. 4. The wonders of whales in Alaska — their life cycles, travels and travails — are examined, with an authoritative history of commercial and subsistence whaling in the North. Includes a fold-out poster of 14 major whale species in Alaska in perspective, color photos and illustrations, with historical photos and line drawings. 144 pages, $12.95.

The Aurora Borealis, Vol. 6, No. 2. The northern lights — in ancient times seen as a dreadful forecast of doom, in modern days an inspiration to countless poets. What causes the aurora, how it works, how and why scientists are studying it today and its implications for our future. 96 pages, $7.95.

Alaska's Native People, Vol. 6, No. 3. Examine the varied worlds of the Inupiat Eskimo, Yup'ik Eskimo, Athabascan, Aleut, Tlingit, Haida and Tsimshian. Included are sensitive, informative articles by Native writers, plus a large, four-color map detailing the Native villages and defining the language areas, 304 pages, $24.95.

The Stikine, Vol. 6, No. 4. River route to three Canadian gold strikes in the 1800s, the Stikine is the largest and most navigable of several rivers that flow from northwestern Canada through southeastern Alaska on their way to the sea. Illustrated with contemporary color photos and historic black-and-white; includes a large fold-out map. 96 pages, $9.95.

Alaska's Great Interior, Vol. 7, No. 1. Alaska's rich Interior country, west from the Alaska-Yukon Territory border and including the huge drainage between the Alaska Range and the Brooks Range, is covered thoroughly. Included are the region's people, communities, history, economy, wilderness areas and wildlife. Illustrated with contemporary color and black-and-white photos. Includes a large fold-out map. 128 pages, $9.95.

A Photographic Geography of Alaska, Vol. 7, No. 2. An overview of the entire state — a visual tour through the six regions of Alaska: Southeast, Southcentral/Gulf Coast, Alaska Peninsula and Aleutians, Bering Sea Coast, Arctic and Interior. Plus a handy appendix of valuable information — "Facts About Alaska." Revised in 1983. Approximately 160 color and black-and-white photos and 35 maps. 192 pages, $15.95.

The Aleutians, Vol. 7, No. 3. Home of the Aleut, a tremendous wildlife spectacle, a major World War II battleground and now the heart of a thriving new commercial fishing industry. Contemporary color and black-and-white photographs, and a large fold-out map. 224 pages, $14.95.

Klondike Lost: A Decade of Photographs by Kinsey & Kinsey, Vol. 7, No. 4. An album of rare photographs and all-new text about the lost Klondike boom town of Grand Forks, second in size only to Dawson during the gold rush. $12.95.

Wrangell-Saint Elias, Vol. 8, No. 1. Mountains, including the continent's second- and fourth-highest peaks, dominate this international wilderness that sweeps from the Wrangell Mountains in Alaska to the southern Saint Elias range in Canada. Includes a large fold-out map. 144 pages, $9.95.

Alaska Mammals, Vol. 8, No. 2. From tiny ground squirrels to the powerful polar bear, and from the tundra to the magnificent whales inhabiting Alaska's waters, this volume includes 80 species of mammals found in Alaska. 184 pages, $12.95.

The Kotzebue Basin, Vol. 8, No. 3. Examines northwestern Alaska's thriving trading area of Kotzebue Sound and the Kobuk and Noatak river basins, lifelines of the region's Inupiat Eskimos, early explorers, and present-day, hardy residents. 184 pages, $12.95.

Alaska National Interest Lands, Vol. 8, No. 4. Following passage of the bill formalizing Alaska's national interest land selections (d-2 lands), longtime Alaskans Celia Hunter and Ginny Wood review each selection, outlining location, size, access, and briefly describing the region's special attractions. 242 pages, $14.95.

Alaska's Glaciers, Vol. 9, No. 1. Examines in depth the massive rivers of ice, their composition, exploration, present-day distribution and scientific significance. 144 pages, $10.95.

Sitka and Its Ocean/Island World, Vol. 9, No. 2. From the elegant capital of Russian America to a beautiful but modern port, Sitka, on Baranof Island, has become a commercial and cultural center for southeastern Alaska. 128 pages, $9.95.

Islands of the Seals: The Pribilofs, Vol. 9, No. 3. Great herds of northern fur seals drew Russians and Aleuts to these remote Bering Sea islands where they founded permanent communities and established a unique international commerce. 128 pages, $9.95.

Alaska's Oil/Gas & Minerals Industry, Vol. 9, No. 4. Experts detail the geological processes and resulting mineral and fossil fuel resources that are now in the forefront of Alaska's economy. Illustrated with historical black-and-white and contemporary color photographs. 216 pages, $12.95.

Adventure Roads North: The Story of the Alaska Highway and Other Roads in *The MILEPOST®*, Vol. 10, No. 1. From Alaska's first highway — the Richardson — to the famous Alaska Highway, first overland route to the 49th state, text and photos provide a history of Alaska's roads and take a mile-by-mile look at the country they cross. 224 pages, $14.95.

ANCHORAGE and the Cook Inlet Basin, Vol. 10, No. 2. "Anchorage country" . . . the Kenai, the Susitna Valley, and Matanuska. Heavily illustrated in color and including three illustrated maps . . . one an uproarious artist's forecast of "Anchorage 2035." 168 pages, $14.95.

Alaska's Salmon Fisheries, Vol. 10, No. 3. The work of *ALASKA®* magazine Outdoors Editor Jim Rearden, this issue takes a comprehensive look at Alaska's most valuable commercial fishery. 128 pages, $12.95.

Up the Koyukuk, Vol. 10, No. 4. Highlights the Koyukuk region of north-central Alaska . . . the wildlife, fauna, Native culture and more. 152 pages. $14.95.

Nome: City of the Golden Beaches, Vol. 11, No. 1. The colorful history of Alaska's most famous gold rush town has never been told like this before. Illustrated with hundreds of rare black-and-white photos, the book traces the story of Nome from the crazy days of the 1900 gold rush. 184 pages, $14.95.

Alaska's Farms and Gardens, Vol. 11, No. 2. An overview of the past, present, and future of agriculture in Alaska, and a wealth of information on how to grow your own fruit and vegetables in the north. 144 pages, $12.95.

Chilkat River Valley, Vol. 11, No. 3. This issue explores the mountain-rimmed valley at the head of the Inside Passage, its natural resources, and those hardy residents who make their home along the Chilkat. 112 pages, $12.95.

Alaska Steam, Vol. 11, No. 4. A pictorial history of the Alaska Steamship Company pioneering the northern travel lanes. Compiled by Lucile McDonald. More than 100 black-and-white historical photos. 160 pages, $12.95.

Northwest Territories, Vol. 12, No. 1. An in-depth look at some of the most beautiful and isolated land in North America. Compiled by Richard Harrington. 148 color photos. 136 pages. $12.95.

Alaska's Forest Resources, Vol. 12, No. 2 examines the majestic and valuable forests of Alaska. Nearly 200 historical black-and-white and color photos. $14.95.

Alaska Native Arts and Crafts, Vol. 12, No. 3. An in-depth look at the art and artifacts of Alaska's Native people. More than 200 full color photos. $17.95.

NEXT ISSUE:
Where Mountains Meet the Sea: Alaska's Gulf Coast, Vol. 13, No. 1. Alaskan's first-hand descriptions of the 850-mile arc that crowns the Pacific Ocean from Kodiak and surrounding islands to Cape Yakataga. Included is a historical overview of this area, and a close look at the geological forces that constantly reshape its landscape. More than 300 photos.

Your $30 membership in the Alaska Geographic Society includes 4 subsequent issues of *ALASKA GEOGRAPHIC®*, the Society's official quarterly. Please add $4 for non-U.S. membership.

Additional membership information available upon request. Single copies of the *ALASKA GEOGRAPHIC®* back issues are also available. When ordering, please make payments in U.S. funds and add $1 postage/handling per copy. To order back issues send your check or money order and volumes desired to:

The Alaska Geographic Society

Box 4 EEE, Anchorage, Alaska 99509